Lecture Notes in Mathematics

Edited by A. Dold and B. Eckmann

1288

Yuri L. Rodin

Generalized Analytic Functions on Riemann Surfaces

Springer-Verlag
Berlin Heidelberg New York London Paris Tokyo

Author

Yuri L. Rodin
Academy of Sciences of the USSR, Institute of Solid State Physics, Chernogolovka,
Moscow Distr. 142432, USSR

Mathematics Subject Classification (1980): 30F30, 30G20

ISBN 3-540-18572-0 Springer-Verlag Berlin Heidelberg New York
ISBN 0-387-18572-0 Springer-Verlag New York Berlin Heidelberg

Printed in Germany

Printing and binding: Druckhaus Beltz, Hemsbach/Bergstr.
2146/3140-543210

PREFACE

This book presents results arising from several areas of the theory of functions and mathematical physics.

The first of these sources, the theory of generalized analytic (pseudo-analytic) functions of L. Bers [a,b] and I.N. Vekua [a,b] has been constructed within the framework of a general interest in different generalizations of analyticity. It was established that such fundamental properties of analytic functions as the argument principle, the Liouville theorem and so on are inherent in solutions of all linear elliptic systems of first order with two unknown functions on the plane. By quasiconformal mappings these systems can be reduced to the complex Carleman-Bers-Vekua equation

$$\bar{\partial} u + au + b\bar{u} = 0 \tag{1}$$

Later the theory of matrix equations (1) was built (W. Wendland [a]). These equations are extremely important for applications (see §12).

At the same time on Riemann surfaces the Riemann boundary problem

$$F^{+}(p) = G(p)\ F^{-}(p) \tag{2}$$

was studied (A. Grothendieck [a], W. Koppelman [b,c], Yu.L. Rodin [a,c,p], H. Röhrl [a,b] and other authors). Main facts of the algebraic function theory were related with the theory of singular integral operators and the classification problem of vector bundles over Riemann surfaces. Afterwards this theory found fundamental physical applications (the Riemann problem method of V.E. Zakharov - A.B. Shabat) in the inverse scattering problem, the integrable systems theory and the solitons theory. At last, recently generalized analytic functions were used in these areas too (see M.J. Ablowitz, D. Bar Yaacov, A.S. Fokas [a], A.S. Fokas, M.J. Ablowitz [a,b], I.M. Krichever, S.P. Novikov [a], A.V. Mikhailov [a,b], V.E. Zakharov, S.V. Manakov [a], V.E. Zakharov, A.V. Mikhailov [a]).

These circumstance stimulated the study of generalized analytic functions on Riemann surfaces. The work was begun by L. Bers [c] and was continued by W. Koppelman [c] and the author [d-j,l]. In our book this area is presented systematically for the first time.

Chapter 1 is devoted to the Riemann-Roch theorem and, naturally, is enclosed into the general theory of the index of elliptic operators with corresponding simplifications. In Chapter 2 multivalued solutions of equation (1) are studied. It demands to look for some representations of them. In particular, the methods allow to obtain a direct proof of the Riemann-Roch theorem. In Chapter 5 they are used to study singular cases and surfaces of infinite genus. In Chapter 3 we expound the Riemann boundary problem and its connections with the Riemann-Roch and the Abel theorems, the Jacobi inversion problem and the classification problem for bundles. The main and most difficult problem of generalized analytic function theory is solved in Chapter 4. It is known that the Abel problem of the existence of an analytic function with prescribed zeros and poles on a compact Riemann surface cannot be solved by pure algebraic methods and demands to use a transcendental operation - applying the logarithm. In our case it leads to a nonlinear integral equation. This equation has been a success to investigate the problem completely.

At last, in §12 we describe very briefly some approaches to physical applications. This is a subject of the expository paper of the author which was published in the journal "Physica D" recently.

The book is addressed to mathematicians and physicists, specialists in the theory of functions, differential equations and mathematical physics (field theory, solitions theory and so on). A preliminary knowledge of the theory of Riemann surfaces and algebraic topology is not necessary for reading the book.

The author is sincerely thankful to his tutor professor L.I. Volkoviskii. He was the initiator for the study of the Riemann problem and generalized analytic functions on Riemann surfaces in the USSR and directed the author's work during many years. The author is glad to express his gratitude to Prof. V.E. Zakharov and Prof. A.V. Mikhailov for numerous fruitful discussions on physical applications of Riemann surfaces. The author is also grateful to Prof. Dr. W.L. Wendland whose moral support was decisive and to Prof. Dr. H. Begehr who edited the manuscript and inserted a number of improvements.

C O N T E N T S

CHAPTER 1

THE DOLBEAULT AND RIEMANN-ROCH THEOREMS

§ 1. Generalized analytic functions in the disk

A. The operator T

1. Consider the Cauchy-Riemann equations

$$\begin{pmatrix} \frac{\partial}{\partial x} & -\frac{\partial}{\partial y} \\ \frac{\partial}{\partial y} & \frac{\partial}{\partial x} \end{pmatrix} \begin{pmatrix} \varphi \\ \psi \end{pmatrix} = 0 . \tag{1.1}$$

Letting $u = \varphi + i\psi$ and introducing the operators of complex differentiation

$$\bar{\partial} = \frac{\partial}{\partial \bar{z}} = \frac{1}{2}\left(\frac{\partial}{\partial x} + i\frac{\partial}{\partial y}\right), \quad \partial = \frac{\partial}{\partial z} = \frac{1}{2}\left(\frac{\partial}{\partial x} - i\frac{\partial}{\partial y}\right), \tag{1.2}$$

$$z = x + iy$$

we rewrite equation (1.1) in the form

$$\bar{\partial} u = 0 . \tag{1.3}$$

The corresponding inhomogeneous equation has the form

$$\bar{\partial} u = f . \tag{1.4}$$

Let G be a bounded domain of the complex z-plane with a sufficiently smooth boundary ∂G, $\bar{G}$ be its closure and f be a function continuous in G. Then the general solution of (1.4) has the form

$$u(z) = F(z) - \frac{1}{\pi} \iint_G \frac{f(t)\, d\sigma_t}{t-z} , \tag{1.5}$$

$$d\sigma_t = d\xi d\eta , \quad t = \xi + i\eta .$$

Here $F(z)$ is an arbitrary analytic function in G.

We use the Green formulae in the form

$$\iint_G \frac{\partial g}{\partial \bar{z}}\, d\sigma_z = \frac{1}{2i} \int_{\partial G} g dz ,$$

$$\iint_G \frac{\partial g}{\partial z}\, d\sigma_z = - \frac{1}{2i} \int_{\partial G} g d\bar{z} . \tag{1.6}$$

Then, for any function of the class C^1 in the closed domain $\bar{G}$ the well-known formula

$$u(z) = - \frac{1}{\pi} \iint_G \frac{\partial u}{\partial \bar{t}} \frac{d\sigma_t}{t-z} + \frac{1}{2\pi i} \int_{\partial G} \frac{u(\tau) d\tau}{\tau - z} \tag{1.7}$$

is valid. Equation (1.7) involves (1.5).

2. Below, (1.4) will be considered for more weak assumptions. In order to make sure all these formulae are valid for wider function classes, we describe properties of the operator

$$Tf(z) = - \frac{1}{\pi} \iint_G \frac{f(t) d\sigma_t}{t-z} . \tag{1.8}$$

This operator belongs to the class of operators of the potential type (A. Calderon, A. Zygmund [a]). We list the properties of the operator T following I.N. Vekua [a].

First some Banach spaces are introduced which will be used below. Let a function $f(z)$ satisfy the Hölder condition

$$|f(z_1) - f(z_2)| \leq H |z_1 - z_2|^\alpha , \qquad 0 < \alpha \leq 1 , \tag{1.9}$$

in the closed domain $\bar{G}$. Denote

$$H(f) = \inf H = \sup_{z_1, z_2 \in \bar{G}} \frac{|f(z_1) - f(z_2)|}{|z_1 - z_2|^\alpha} .$$

Introduce the Banach space $C_\alpha(\bar{G})$ of functions satisfying the Hölder condition with exponent α in $\bar{G}$ with the norm

$$\| f \|_{C_\alpha(\bar{G})} = \max_{z \in \bar{G}} |f(z)| + H(f) = \| f \|_{C(\bar{G})} + H(f) . \tag{1.10}$$

Let $f \in L_p(\bar{G})$, $0 < \alpha \leq 1$, be some constant and

$$B(f) = \sup \frac{1}{|\Delta z|^{\alpha}} \| f(z+\Delta z) - f(z) \|_{L_p(\bar{G})} .$$

Introduce the Banach space $L_{p,\alpha}(\bar{G})$ of functions satisfying the inequality

$$\| f(z+\Delta z) - f(z) \|_{L_p(\bar{G})} \leq B(f) \; |\Delta z|^{\alpha} \tag{1.11}$$

with the norm

$$\| f \|_{L_p^{\alpha}(\bar{G})} = \| f \|_{L_p(\bar{G})} + B(f) . \tag{1.12}$$

The set of functions continuous in $\bar{G}$ together with their partial derivatives up to the order m inclusive forms the Banach space $C_m(\bar{G})$ with the norm

$$\| f \|_{C_m(\bar{G})} = \sum_{k=0}^{m} \sum_{\ell=0}^{k} \max_{z\in\bar{G}} \left| \frac{\partial^k f}{\partial z^{k-\ell} \, \partial \bar{z}^{\ell}} \right| . \tag{1.13}$$

If all partial derivatives satisfy the Hölder condition, we obtain the space $C_{m,\alpha}(\bar{G})$ with the norm

$$\| f \|_{C_{m,\alpha}(\bar{G})} = \sum_{k=0}^{m} \sum_{\ell=0}^{k} \left\{ \left\| \frac{\partial^k f}{\partial z^{k-\ell} \, \partial \bar{z}^{\ell}} \right\|_{C(\bar{G})} + H\left(\frac{\partial^k f}{\partial z^{k-\ell} \, \partial \bar{z}^{\ell}} \right) \right\} . \tag{1.14}$$

Theorem 1.1. Let $f \in L_p(\bar{G})$, $p > 2$, and

$$g(z) = Tf(z) = - \frac{1}{\pi} \iint_G \frac{f(t)\, d\sigma_t}{t-z} .$$

Then the following estimations are valid:

$$|g(z)| \leq M_1 \| f \|_{L_p(\bar{G})} ,$$

$$|g(z_1) - g(z_2)| \leq M_2 \| f \|_{L_p(\bar{G})} |z_1 - z_2|^{\alpha} , \quad \alpha = \frac{p-2}{p} . \tag{1.15}$$

Hence the linear operator

$$T: L_p(\bar{G}) \to C_{\alpha}(\bar{G}) , \; \alpha = \frac{p-2}{p} , \; p > 2 ,$$

is compact and

$$\| Tf \|_{C_\alpha(\bar{G})} \le M \| f \|_{L_p(\bar{G})} \ . \tag{1.16}$$

Theorem 1.2. If $f \in L_p(\bar{G})$, $1 \le p \le 2$, then the function $g(z) = T f(z)$ belongs to the space $L_{\gamma,\alpha}(\bar{G})$, where γ is an arbitrary number satisfying the inequality

$$1 < \gamma < \frac{2p}{2-p} \ . \tag{1.17}$$

Moreover, the following estimations are valid,

$$\| Tf \|_{L_\gamma(\bar{G})} \le M_{p,\gamma} \| f \|_{L_p(\bar{G})} \ ,$$

$$\Big(\iint_G |g(z+\Delta z) - g(z)|^\gamma \, d\sigma_z\Big)^{1/\gamma} \le M'_{p,\gamma} \| f \|_{L_p(\bar{G})} |\Delta z|^\alpha \ , \tag{1.18}$$

$$\alpha = \frac{1}{\gamma} - \frac{2-p}{2p} > 0 \ .$$

This result entails the complete continuity of the operator T mapping

$$T: L_p(\bar{G}) \to L_{\gamma,\alpha}(\bar{G}) \ , \ 1 \le p \le 2 \ , \ \alpha = \frac{1}{\gamma} - \frac{1}{p} + \frac{1}{2} \ , \ p \le \gamma < \frac{2p}{2-p} \ .$$

In the following we understand derivatives in the generalized sense. The linear set of functions belonging to $C_m(\bar{G})$ and having compact support in G is denoted by $C^0_m(G)$.

Definition. Let $f,g \in L_1(G)$ and satisfy the relation

$$\iint_G g \frac{\partial\varphi}{\partial\bar{z}} \, d\sigma_z + \iint_G f \varphi \, d\sigma_z = 0 \tag{1.19}$$

$$\Big(\iint_G g \frac{\partial\varphi}{\partial z} \, d\sigma_z + \iint_G f \varphi \, d\sigma_z = 0 \ , \tag{1.19'}$$

respectively$\Big)$ for an arbitrary function $\varphi \in C^0_1(G)$. Then the function f is said to be the generalized derivative of g with respect to $\bar{z}$ (with respect to z , respectively)

$$f = \frac{\partial g}{\partial\bar{z}} \qquad \Big(f = \frac{\partial g}{\partial z}\Big) \ .$$

The class of functions possessing generalized derivatives with respect to $\bar{z}$ is denoted by $D_{\bar{z}}(G)$ ($D_z(G)$, respectively).

The class of functions possessing generalized derivatives belonging to L_p is denoted by $D_{1,p}(G)$. The Banach space of functions possessing generalized derivatives of order $\leq m$ with the norm

$$\| f \|_{D_{m,p}(G)} = \sum_{\ell,k=0}^{\ell+k\leq m} \left\| \frac{\partial^{\ell+k} f}{\partial z^{\ell}\, \partial \bar{z}^{k}} \right\|_{L_p(\bar{G})} \tag{1.20}$$

is denoted by $D_{m,p}(G)$.

In the case the derivatives are integrable in the closed domain $\bar{G}$ we write $D_{m,p}(\bar{G})$.

Theorem 1.3. If $f = \partial_{\bar{z}} g \in L_1(\bar{G})$, then

$$g(z) = \phi(z) - \frac{1}{\pi} \iint_G \frac{f(t)\,d\sigma_t}{t-z} \tag{1.21}$$

where ϕ is a holomorphic function in G. Conversely, if $\phi(z)$ is a holomorphic function in G and $f \in L_1(\bar{G})$, then the function $g(z) = \phi(z) + T f(z) \in D_{\bar{z}}(G)$ and $\bar{\partial} g = f$. If $u(z) \in C(\bar{G})$ and $\partial_{\bar{z}} u \in L_p(\bar{G})$, $p > 2$, then equation (1.7) is valid.

Theorem 1.4. Let $f(z) \in C_{m,\alpha}(\bar{G})$, $0 < \alpha < 1$, $m \geq 0$. Then the function $g(z) = T f(z)$ belongs to the class $C_{m+1,\alpha}(\bar{G})$, the operator T is completely continuous in the space $C_{m,\alpha}(\bar{G})$ and

$$\frac{\partial g}{\partial \bar{z}} = f, \quad \frac{\partial g}{\partial z} = \Pi_f = -\frac{1}{\pi} \iint_G \frac{f(t)\,d\sigma_t}{(t-z)^2}. \tag{1.22}$$

The integral in (1.22) is understood in the sense of the prinicipal value. The operator Π is a linear bounded operator in $C_{m,\alpha}(\bar{G})$ mapping this space into itself. The operator Π can be continued up to a unitary operator in $L_2(G)$ and up to a bounded operator in any $L_p(G)$, $p > 1$. The first formula (1.22) is valid also for $f \in L_1(\bar{G})$.

Consider the operator

$$Pf = \frac{1}{\pi} \iint_G \frac{a(t) f(t)\,d\sigma_t}{t-z}. \tag{1.23}$$

Theorem 1.5. Let $a(z) \in L_p(\bar{G})$, $p > 2$. Then the operator (1.23) is completely continuous in the space $C(\bar{G})$, maps this space into $C_\alpha(\bar{G})$, $\alpha = \frac{p-2}{p}$ and

$$\| Pf \|_{C_\alpha(\bar{G})} \le M_p \ \| a \|_{L_p(\bar{G})} \ \| f \|_{C(\bar{G})} .$$

Moreover, this operator is completely continuous in the space $L_q(\bar{G})$, $\frac{1}{2} \le \frac{1}{p} + \frac{1}{q} \le 1$, too. If an integer n satisfies the condition

$$n - 1 \le \frac{2p}{p-2} \left(\frac{1}{p} + \frac{1}{q} - \frac{1}{2}\right) < n ,$$

then

$$\| P^k f \|_{L^\alpha_{\gamma_k}(\mathbb{C})} \le M_{p,q,\alpha} \ \| a \|_{L_p(\bar{G})} \ \| f \|_{L_q(\bar{G})} ,$$

$$k = 1,\ldots,n ,$$

$$\| P^{n+1} f \|_{C_\beta(\mathbb{C})} \le M'_{p,q,\alpha} \ \| a \|_{L_p(\bar{G})} \ \| f \|_{L_q(\bar{G})} ,$$

where

$$\frac{1}{\gamma_k} = \frac{1}{q} + \frac{k}{p} - \frac{k}{2} + k\alpha , \ k = 1,\ldots,n ,$$

$$\beta = 1 - 2 \left(\frac{1}{q} + \frac{n+1}{p} + n\alpha\right) + n ,$$

and α is an arbitrary number satisfying the inequality

$$0 < \alpha < \frac{p-2}{2p} - \frac{1}{n} \left(\frac{1}{p} + \frac{1}{q} - \frac{1}{2}\right) .$$

The reader may find the proofs of these facts and related ones in the book I.N. Vekua [a] .

B. The Carleman-Bers-Vekua system

1. Obviously, the elliptic system

$$\bar{\partial} u + au = 0 \tag{1.24}$$

is reduced to the inhomogeneous Cauchy-Riemann equation (1.4) by taking the logarithm. Equation (1.4) entails the representation for the general solution of (1.24) in the bounded domain G

$$u(z) = \varphi(z)\exp \frac{1}{\pi} \iint\limits_{G} a(t)\, \frac{d\sigma_t}{t-z}\ .$$

Here $\varphi(z)$ is an arbitrary analytic function in G . In particular, all zeros and poles of the function $u(z)$ are determined by the multiplier $\varphi(z)$. It provides a natural way to define orders of zeros and poles and to generalize the argument principle.

A more general system than (1.24) is the Carleman-Bers-Vekua (CBV) system

$$\underset{\sim}{\partial}u \equiv \bar{\partial}u + au + b\bar{u} = 0\ . \tag{1.25}$$

As a rule, we assume that $a,b \in L_p(\bar{G})$, $p > 2$.

The function $u(z)$ is called a solution of the equation (1.25) in the vicinity G_0 of the point z_0 if $u \in D_{\bar{z}}(G_0)$ and the equation (1.25) is valid almost everywhere in G_0 . If $u(z)$ is a solution of (1.25) in the vicinity of every point of the domain G , $u(z)$ is called a regular solution of (1.25). If $u(z)$ is a solution of (1.25) in the vicinity of every point of the domain G except some discrete set of points $G^* \subset G$, called singularities, then following I.N. Vekua [a] such a solution is called a generalized solution. Generalized and regular solutions of the inhomogeneous equation $\underset{\sim}{\partial}u = F$, $F \in L_p(\bar{G})$, $p > 2$, are defined in an analogous manner.

By Theorem 1.3 the class of generalized solutions of the Cauchy-Riemann equation $\bar{\partial}u = 0$ coincides with the class $A^*(G)$ of analytic functions in the domain G with singularities at the points of G^* ; the class of regular solutions of the Cauchy-Riemann equation coincides with the class $A(G)$ of functions holomorphic in the domain G .

We denote by $\tilde{A}^*(a,b,F,G)$ the class of generalized solutions of (1.25) such that $\bar{\partial}u = -au - b\bar{u} + F \in L_1(G)$.
It is clear that this class contains the class $\tilde{A}(a,b,F,G)$ of regular solutions and solutions with singularities of order less than two if the coefficients of the equation at the points of G are bounded. If $a,\ b,\ F \in L_p(G)$, we write $\tilde{A}^*_p(a,b,F,G)$ and $\tilde{A}_p(a,b,F,G)$, respectively. The union of all classes $\tilde{A}^*_p(a,b,F,G)$ corresponding to all $a,\ b,\ F$ for fixed p is denoted by $\tilde{A}^*_p(G)$ (and $\tilde{A}_p(G)$, respectively). For $F \equiv 0$ we write $A^*_p(a,b,G)$, $A^*_p(G)$, $A_p(a,b,G)$, $A_p(G)$. All these notations are due to I.N. Vekua [a].

By Theorem 1.3 these solutions are representable in the form

$$u - Pu = \phi(z) + TF \tag{1.26}$$

where

$$Pf = - T(af + b\bar{f}) \ , \ Tf = - \frac{1}{\pi} \iint_G \frac{f(t)\,d\sigma_t}{t-z} \ , \tag{1.27}$$

and $\phi(z)$ is a holomorphic function in G. For $F \equiv 0$ we obtain the integral equation

$$u(z) - \frac{1}{\pi} \iint_G [a(t)\ u(t) + b(t)\ \overline{u(t)}]\ \frac{d\sigma_t}{t-z} = \phi(z) \tag{1.28}$$

for generalized analytic functions.

Let $a, b, F \in L_p(\bar{G})$, $p > 2$, and the function $u(z)$ in (1.26) be continuous in G. Then, by Theorem 1.1, the functions Pu and TF belong to the class $C_\alpha(\bar{G})$, $\alpha = \frac{p-2}{p}$, are analytic in the domain $\mathbb{C} - \bar{G}$ ($\mathbb{C}$ is the complex plane) and are equal to zero at infinity. It entails the representation

$$\phi(z) = \frac{1}{2\pi i} \int_{\partial G} \frac{u(t)\,dt}{t-z} \tag{1.29}$$

<u>Theorem 1.6</u>. If $u(z)$ is a regular solution of the equation $\underset{\sim}{\partial} u = F$, $a, b, F \in L_p(\bar{G})$, $p > 2$, $u \in \tilde{A}_p(a,b,F,G)$, then $u(z)$ satisfies the Hölder condition, $u \in C_\alpha(\bar{G})$, $\alpha = \frac{p-2}{2}$.

For the proof see I.N. Vekua [a].

2. The following theorem is called the Bers-Vekua similarity principle.

<u>Theorem 1.7</u>. Let $u(z)$ be a generalized solution of (1.25), $u \in A_p^*(a,b,G)$, $p > 2$, and

$$g(z) = \begin{cases} a(z)+b(z)\ \dfrac{\overline{u(z)}}{u(z)} \ , & z \in G \setminus \{G^* \cup \{z:\ u(z) = 0\}\} \\ a(z)+b(z) \ , & z \in G^* \cup \{z:\ u(z) = 0\}\ . \end{cases} \tag{1.30}$$

Then the function

$$\varphi(z) = u(z) \exp \{- \frac{1}{\pi} \iint_G g(t)\ \frac{d\sigma_t}{t-z}\} \tag{1.31}$$

is analytic in the domain $G \setminus G^*$, $\varphi \in A^*(G)$.

Since $g \in L_p(G)$, $p > 2$, the right hand side of (1.31) belongs to $D_{\bar{z}}(G \setminus G^*)$ where G^* is the singularities set of the solution $u(z)$.

Then

$$\bar{\partial}\varphi = \{u(z)g(z) - au - b\bar{u}\}\exp\{-\frac{1}{\pi}\iint_G g(t)\frac{d\sigma_t}{t-z}\} = 0 .$$

almost everywhere in $G \setminus G^*$. This entails the holomorphy of φ in $G \setminus G^*$. In particular, if $u(z)$ is a regular solution, then the function $\varphi(z)$ is holomorphic in G.

Formula (1.31) involves some consequences. The most important one is the argument principle: the difference between the numbers of zeros and poles (taking their orders into account) of a generalized analytic function in the domain G is equal to

$$N_G - P_G = \frac{1}{2\pi}\Delta_{\partial G}\arg u(z) . \qquad (1.32)$$

The formulae (1.28), (1.31) were obtained by N. Theodoresco [a,b], Carleman [a,b], L. Bers [a,b], I.N. Vekua [a,b].

3. Let us return to the integral equation

$$u - Pu \equiv u(z) - \frac{1}{\pi}\iint_G (a(t)u(t) + b(t)\overline{u(t)})\frac{d\sigma_t}{t-z} = g(z) \qquad (1.33)$$

for the case $a,b \in L_p(\bar{G})$, $p > 2$ and show that it is solvable for any right hand side $g \in L_q(\bar{G})$, $q \geq \frac{p}{p-1}$.

By Theorem 1.5 the operator Pu is completely continuous in the space $L_q(\bar{G})$, $q \geq \frac{p}{p-1}$. Therefore, it is sufficient to show that the homogeneous equation

$$u - Pu = 0$$

has no nontrivial solutions.

Let $u_0 \in L_q(\bar{G})$, $q \geq \frac{p}{p-1}$, be a solution of equation (1.33). Then $u_0 = Pu_0 = \ldots = P^n u_0$. By Theorem 1.5 there exists such an n for which $P^n u_0 \in C_\alpha(G)$. Hence u_0 is continuous in $\bar{G}$ and satisfies a Hölder condition. By Theorem 1.4 the function $u_0 = Pu_0$ belongs to the class $D_{\bar{z}}(\bar{G})$ and, consequently, is a regular solution of the equation

$$\underset{\sim}{\partial} u \equiv \bar{\partial}u + au + b\bar{u} = 0 .$$

By formula (1.32) the value $\frac{1}{2\pi}\Delta_{\partial G}\arg u_0(z)$ is non-negative and equal to the sum of the orders of zeros of the function u_0 in the domain G. On the other hand, the function

$$u_0(z) - Pu_0(z) \equiv u_0(z) - \frac{1}{\pi}\iint_G (a(t)u_0(t) + b(t)\overline{u_0(t)}) \frac{d\sigma_t}{t-z}$$

is holomorphic in the domain $\mathbb{C}\setminus\bar{G}$ and is equal to zero at infinity. This means that $\frac{1}{2\pi}\arg_{\partial G} u_0(z) \le -1$ if $u_0 \not\equiv 0$. This contradiction proves that $u_0 \equiv 0$.

Therefore, any generalized analytic function in the domain G having poles which orders ≤ 1 is a solution of the integral equation

$$u - Pu \equiv u(z) - \frac{1}{\pi}\iint_G (a(t)u(t) + b(t)\overline{u(t)}) \frac{d\sigma_t}{t-z} = \phi(z) \qquad (1.34)$$

where the analytic function $\phi(z)$ and $u(z)$ have the same poles. Conversely, if $\phi(z)$ is an analytic function in G continuous in $\bar{G}$ up to poles of first order, then $\phi \in L_q(\bar{G})$, $\frac{p}{p-1} \le q < 2$. In this case (1.33) has a solution $u \in L_q(\bar{G})$ being a generalized analytic function in G with poles determined by the function $\phi(z)$.

4. Theorem 1.8. (Poincaré lemma). Let $a, b, F \in L_p(\bar{G})$, $p > 2$. Then the equation $\underset{\sim}{\partial} u = F$ is solvable in any space $L_q(\bar{G})$, $q > \frac{p}{p-1}$.

Consider the equation

$$u - Pu = TF \qquad (1.35)$$

The function $TF \in C_\alpha(G)$, $\alpha = \frac{p-2}{p}$ by Theorem 1.1. As it was shown above, the equation $u - Pu = 0$ has no non-trivial solutions in $C_\alpha(\bar{G})$. It entails the unique solvability of equation (1.35). It is clear, that the solution of (1.35) is a function of the class $\mathfrak{A}(a,b,F,G)$ and hence is a regular solution of the equation $\underset{\sim}{\partial} u = F$.

§ 2. The Carleman-Bers-Vekua System on Riemann surfaces

A. Riemann surfaces

Let M be a closed Riemann surface of genus g. As it is known, a Riemann surface is a topological Hausdorff space with a complex structure. The complex structure is determined by the set of simply-connected coordinate neighborhoods U such that to any point $p \in M$ there belongs at least one coordinate neighborhood. In any coordinate neighborhood U one defines the local coordinate $z(p)$, $p \in U$, mapping U into the unit disk $|z| < 1$ of the complex z-plane. If $U \cap U'$ is not empty, the relations between the corresponding local coordinates z and z' are analytic in this set and $z = z(z')$, $z' = z'(z)$ are conformal mappings.

For example, the equation

$$w^2 = (z-z_1)(z-z_2)(z-z_3)(z-z_4) \tag{2.1}$$

determines a two-sheeted surface over the z-plane. It may be constructed by attaching two copies of z-planes cut along lines connecting the points z_1, z_2 and z_3, z_4. As it is seen from Figure 1, this surface is topologically equivalent to a torus.

A compact (closed) Riemann surface is homeomorphic to a sphere with g handles. For $g = 0$ we have a sphere, and for $g = 1$ a torus. A typical property of all surfaces for $g > 0$ is the existence of cyclic sections, i.e. closed curves not separating the surface (see figure 2). For any handle there exist two kinds of such oriented sections (for a torus a parallel and a meridian one).

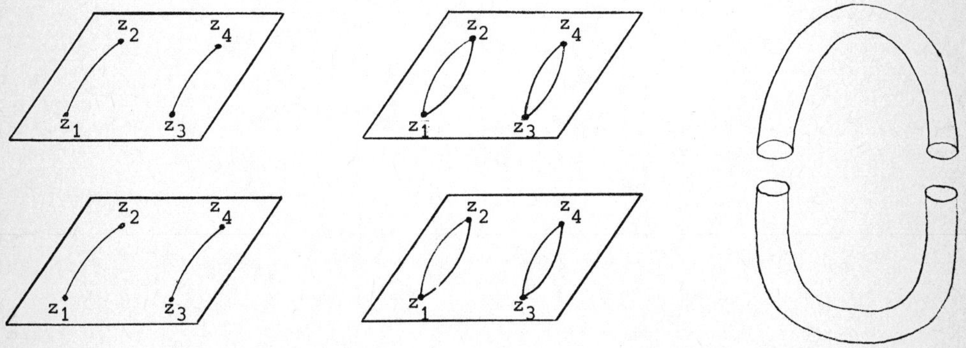

Figure 1

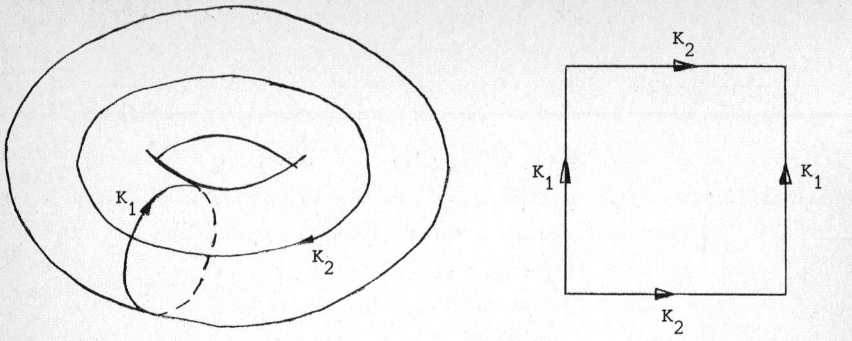

Figure 2

These sections may be numbered in such a manner that every even cycle intersects every odd one from the right to the left (in a topological language it means that the intersection index is equal to $I(K_{2j-1},K_{2j}) = 1$) and intersects no other cycles. Deform $2g$ of these cycles such that any two of them are intersecting in a single point or nowwhere and cut the surface along these cycles (figure 2,3).

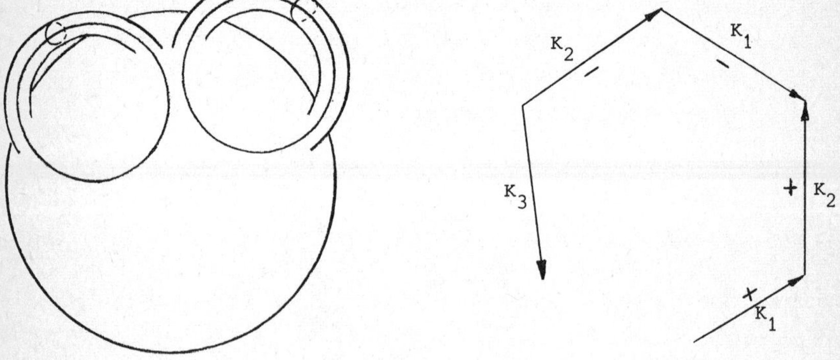

Figure 3

We obtain the 4g-sided polygon with pairs of sides oriented to meet each other. Below we fix these orientations and distinguish the sides of the polygon by the signs "+" and "-" correspondingly to the opposite banks of the cut.

It can be shown that the cycles $K_1,\dots,K_{2g}$ form a basis of the onedimensional homology group $H_1(M)$ (the Betty group). The reader may represent elements of this homology group as linear combinations of the type $\sum_1^{2g} c_j K_j$. Here c_j are elements of the basic ring (real or complex numbers). Such a basis is called canonical. In the following we will concider a fixed canonical basis. The surface M cutted along this basis is donoted by M. The representation of M in the polygon form (figure 3) shows the triangulability of M. A triangulation of the surface is its subdivision into a countable (finite, if the surface is compact) set of closed triangles satisfying the following conditions

a) the triangles have no common interior points

b) two triangles may possess a common side or vertex

c) any vertex belongs to a finite set of triangles.

Below we assume that triangle boundaries are oriented such that the interior of a triangle is located at the left of the boundary cycle. The coordinated choice of the triangles orientation (common boundary sides of triangles are passed in the opposite directions) is possible only for the class of so-called orientable surfaces. Riemann surfaces belong to this class. Below, we also assume that in any triangle $U_j (j = 1,2,\dots)$ of the triangulation there is defined a local coordinate z_j. This triangulation $\mathcal{M} = \{U_j, j \in I\}$, where I is some set of indices, is fixed everywhere in this book.

B. Spaces of functions and differentials

In this section we adduce definitions of basic functional spaces which will be used below.

In any domain U_j of the triangulation $\mathcal{M}$ fixed above we define the spaces $C(\bar{U}_j)$, $C_\alpha(\bar{U}_j)$, $L_p(\bar{U}_j)$, $C_{m,\alpha}(\bar{U}_j)$, $L_{p,\alpha}(\bar{U}_j)$ with norms

$$\| f \|_{C(\bar{U}_j)} = \max_{z_j \in z_j(\bar{U}_j)} |f(z_j)| , \tag{2.2$_1$}$$

$$\| f \|_{C_\alpha(\bar{U}_j)} = \| f \|_{C(\bar{U}_j)} + H_j(f) , \tag{2.2$_2$}$$

$$\| f \|_{L_p(\bar{U}_j)} = \Big(\iint_{U_j} |f(z_j)|^p \, d\sigma_{z_j} \Big)^{1/p} , \tag{2.2$_3$}$$

$$\| f \|_{L_p^\alpha(\bar{U}_j)} = \| f \|_{L_p(\bar{U}_j)} + B_j(f) , \tag{2.2$_4$}$$

$$\| f \|_{C_{m,\alpha}(\bar{U}_j)} = \sum_{k=0}^{m} \sum_{\ell=0}^{k} \left\{ \Big\| \frac{\partial^k f}{\partial z^{k-\ell} \partial \bar{z}^{\ell}} \Big\|_{C(\bar{U}_j)} + H_j \Big(\frac{\partial^k f}{\partial z^{k-\ell} \partial \bar{z}^{\ell}} \Big) \right\} . \tag{2.2$_5$}$$

For $\alpha = 0$

$$\| f \|_{C_{m,0}(\bar{U}_j)} = \sum_{k=0}^{m} \sum_{\ell=0}^{k} \left\{ \Big\| \frac{\partial^k f}{\partial z^{k-\ell} \partial \bar{z}^{\ell}} \Big\|_{C(\bar{U}_j)} \right\} . \tag{2.2$'_5$}$$

Here z_j is the fixed local coordinate in U_j . The constants $H_j(f)$ and $B_j(f)$ are calculated for the function $f(z_j(q))$, $q \in U_j$, in the domain $z_j(U_j)$ of the z_j-plane.

Let G be some damain on the surface M that may coincide with M . Let some subset $M_G = \{U_j, j \in I_G\}$ of the triangulation M , $I_G \subset I$, form a triangulation of the domain G . In the case $G = M$ we have $I_G = I$. Define the spaces of functions in G $C^0(\bar{G})$, $C_\alpha^0(\bar{G})$, $L_p^0(\bar{G})$, $L_{p,\alpha}^0(\bar{G})$, $C_{m,\alpha}^0(\bar{G})$ with norms

$$\| f \|_{X(G)} = \sum_{j \in I_G} \| f \|_{X(U_j)} \tag{2.3}$$

where $X(G)$ are arbitrary spaces listed above and $X(U_j)$ are the corresponding spaces from (2.2).

Or course, the norms (2.3) depend on the choice of the triangulation and the local coordinates. But it is easy to verify that new norms are equivalent to initial ones*) .

Let $f(p)$ be a function of the class $C_{1,0}^0(G)$. Then its differential

$$df = \frac{\partial f(p)}{\partial z(p)} \, dz(p) + \frac{\partial f(p)}{\partial \overline{z(p)}} \, d\overline{z(p)} \tag{2.4}$$

*) The norms $\| \cdot \|$ and $\| \cdot \|_1$ are called equivalent if there exist constants C and C' such that

$$C \| \cdot \|_1 \le \| \cdot \| \le C' \| \cdot \|_1 .$$

is independent on the choice of a local coordinate $z(p)$ of the point p. By analogy with (2.4) we consider forms

$$\omega(p) = \varphi(z(p))\, dz(p) + \psi(z(p))\, d\overline{z(p)} \tag{2.5}$$

invariant on M. It is clear that the invariantness of the form (2.5) means that the values of φ and ψ are changed when the local coordinate z is replaced by z_1 by the rule

$$\varphi(z_1(p)) = \varphi(z(p)) \frac{dz(p)}{dz_1(p)}, \quad \psi(z_1(p)) = \psi(z(p)) \frac{d\overline{z(p)}}{d\overline{z_1(p)}} . \tag{2.6}$$

The values of φ and ψ depending on a local coordinate by the law (2.6) are called covariants of the type $(1,0)$ and $(0,1)$, respectively (or covariants with respect to z and $\bar{z}$). If $\psi \equiv 0$ ($\varphi \equiv 0$), the form (2.5) is called a form of the type $(1,0)$ ($(0,1)$ respectively).

We consider also covariants $h(z(p))$ of the type $(1,1)$ depending on a local coordinate by the law

$$h(z_1(p)) = h(z(p)) \frac{d\overline{z(p)} \wedge dz(p)}{d\overline{z_1(p)} \wedge dz_1(p)} = h(z(p)) \frac{J(x,y)}{J(x_1,y_1)}$$

$d\bar{z} \wedge dz = 2i\, dx\, dy$, $z = x + iy$,

where $\frac{J(x,y)}{J(x_1,y_1)}$ is the Jacobian of the corresponding mapping.

The formulae (2.2_1) - (2.2_5) and (2.3) define also the space $C^{(i,j)}(\bar{G})$, $C_\alpha^{(i,j)}(\bar{G})$, $L_p^{(i,j)}(\bar{G})$, $L_{p,\alpha}^{(i,j)}(\bar{G})$, $C_{m,\alpha}^{(i,j)}(\bar{G})$ of covariants of the type (i,j) $(i,j = 0,1)$. Replacements of local coordinates lead to replacements of norms by equivalent ones.

C. Generalized analytic functions

Let $G \subset M$ be a domain (G may coincide with M), $a(p)$, $b(p)$ be covariants of the type $(0,1)$ belonging to the space $L_p^{(0,1)}(\bar{G})$, $p > 2$, $\bar{\partial} = \frac{\partial}{\partial \overline{z(p)}}$ be the operator of the differentiation with respect to a local coordinate. Then the operator

$$\underset{\sim}{\partial} u \equiv \bar{\partial} u + au + b\bar{u} \tag{2.7}$$

transforms functions into covariants of the type $(0,1)$ and the Carleman-Bers-Vekua equation

$$\underset{\sim}{\partial} u \equiv \bar{\partial} u + au + b\bar{u} = 0 \tag{2.8}$$

is invariant with respect to a change of local coordinates. The definitions of regular and generalized solutions as given in § 1 keep their meanings.

We keep the notations $\tilde{A}^*(a,b,F,G)$, $\tilde{A}(a,b,F,G)$, $A^*(a,b,G)$, $A(a,b,G)$, $A^*(G)$, $A(G)$, $A_p^*(G)$, $A_p(G)$.

Simultaneously with the operator $\underset{\sim}{\partial}$ we consider the operator

$$\underset{\sim}{\partial}^* v \equiv - \bar{\partial} v + av + \overline{bv} \tag{2.9}$$

defined on covariants of the type (1,0). The operator $\underset{\sim}{\partial}^*$ transforms covariants of the type (1,0) into covariants $w(z(p))$ of the type (1,1) depending on the local coordinate $z(p)$ by the rule (2.6). Corresponding classes of covariants of the type (1,0) are denoted by $\tilde{A}^{*1}(a,b,F,G)$, $\tilde{A}^1(a,b,F,G)$ and so on.

The equation

$$\underset{\sim}{\partial}^* v \equiv - \bar{\partial} v + av + \overline{bv} = 0 \tag{2.10}$$

is defined in G invariantly.

§ 3. The Dolbeault theorem

A. The Dolbeault theorem

Denote the sheaf of germs*) of regular solutions of the equation $\underset{\sim}{\partial} u = 0$ in the domain $G \subset M$ by $Q(\underset{\sim}{\partial}, G)$ and consider the mapping

$$\underset{\sim}{\partial}: \hat{A}^0(G) \to \Lambda^{0,1}(G) ,$$

where $\hat{A}^0(G)$ is the sheaf of germs of the functions of the class $\tilde{\mathcal{A}}_p(G)$ and $A^{0,1}(G)$ is the sheaf of germs of covariants of the type $(0,1)$ of the class $L_p^{0,1}(\bar{G})$, $p > 2$. Here the class $\tilde{\mathcal{A}}_p(G)$ is defined as the set of functions belonging to the class $\tilde{\mathcal{A}}_p(U)$ in every simply connected subdomain $U \subset G$.

It is clear that $Q(\underset{\sim}{\partial}, G) = \operatorname{Ker} \underset{\sim}{\partial}$. The mapping $\underset{\sim}{\partial}$ is an epimorphism, $\operatorname{Im} \underset{\sim}{\partial} = A^{0,1}(G)$. This follows from the Poincaré lemma (Theorem 1.8). Therefore, the sequence

$$0 \to Q(\underset{\sim}{\partial}, G) \xrightarrow{i} \hat{A}^0(G) \xrightarrow{\underset{\sim}{\partial}} A^{0,1}(G) \to 0 \tag{3.1}$$

is exact. It yields the exactness of the sequence of the cohomology groups

$$0 \to H^0(Q(\underset{\sim}{\partial}, G)) \xrightarrow{i^*} H^0(\hat{A}^0(G)) \xrightarrow{\underset{\sim}{\partial}} H^0(A^{0,1}(G)) \xrightarrow{\delta}$$

$$\to H^1(Q(\underset{\sim}{\partial}, G)) \to H^1(\hat{A}^0(G)) \to \dots . \tag{3.2}$$

The sheaf $\hat{A}^0(G)$ is thin and hence all groups $H^k(\hat{A}^0(G))$, $k > 0$, are equal to zero. To show this directly fix some covering $N = \{U_j, j \in J\}$ of M by simply connected coordinate domains U_j, $j \in J$, where J is an indices set. Let $N_G = \{U_j, j \in J_G\}$ be some subset of N, $J_G \subset J$, forming a covering of the domain G. We assume that the triangulation M fixed above is a subset of N. Let $\sum_k \alpha_k(p) \equiv 1$ be an infinite differentiable partition of unity associated with N_G. Let $f = \{f_{ij}\}$ be a 1-cocycle with values in $\hat{A}^0(G)$, $f \in Z^1(\hat{A}^0(G))$. Assume

$$f_j = \sum_{k \in J_G} \alpha_k f_{jk} .$$

*) Main definitions related to sheaves, cohomology groups, exact sequences are listed in the Appendix.

It is clear that

$$f_i - f_j = \sum_{k\in J_G} \alpha_k(f_{ik} - f_{jk}) = f_{ij} \sum_{k\in J_G} \alpha_k = f_{ij} .$$

Therefore, $\{f_{ij}\} = \delta\{f_j\}$. The function f_j , as it is easy to see, belongs to the class $\mathfrak{A}_p(G)$. Hence the cochain $\{f_j\} \in Z^0(\hat{A}^0(G))$. Analogously, one can show that the sheaf $A^{0,1}(G)$ is thin, too. Indeed, if f_{ik} is a covariant of the type (0,1) then $\varphi_k f_{ik}$ is a covariant of the same type.

From Theorem A.1 the exactness of the cohomology sequences

$$0 \to H^0(Q(\underset{\sim}{\partial},G)) \xrightarrow{i^*} H^0(\hat{A}^0(G)) \xrightarrow{\underset{\sim}{\partial}} H^0(A^{0,1}(G)) \to H^1(Q(\underset{\sim}{\partial},G)) \to 0 ,$$

$$0 \to H^k(Q(\underset{\sim}{\partial},G)) \to 0 , \ k > 1 , \tag{3.3}$$

follow. We conclude that

$$H^1(Q(\underset{\sim}{\partial},G)) = H^0(A^{0,1}(G))/\underset{\sim}{\partial}H^0(\hat{A}^0(G)) , \tag{3.4}$$

$$H^k(Q(\underset{\sim}{\partial},G)) = 0 , \ k > 1 . \tag{3.5}$$

It is naturally to call equation (3.4) the Dolbeault theorem for the operator $\underset{\sim}{\partial}$ (compare Theorem A.2).

Designate the sheaf of germs of regular solutions of the equation $\underset{\sim}{\partial}^*v = 0$ in the domain $G \subset M$ (remember, that these are covariants of the type (1,0)) by $Q^1(\underset{\sim}{\partial}^*,G)$ and consider the mapping

$$\underset{\sim}{\partial}^*: \hat{A}^{1,0}(G) \to A^{1,1}(G)$$

where $\hat{A}^{1,0}(G)$ is the sheaf of germs of covariants of the type (1,0) of the class $\mathfrak{A}^1_p(G)$ and $A^{1,1}(G)$ is the sheaf of germs of covariants of the type (1,1) of the class $L_p^{1,1}$, $p > 2$.

The Poincaré lemma remains valid since locally, for a fixed local coordinate Theorem 1.8 is true. If g is a covariant of type (1,1), the equation $\underset{\sim}{\partial}^*v = g$ is invariant under the change of local coordinates. The sheaves $\hat{A}^{1,0}(G)$ and $A^{1,1}(G)$ are thin. We have the exact sequence

$$0 \to Q^1(\underset{\sim}{\partial},G) \xrightarrow{i} \hat{A}^{1,0}(G) \xrightarrow{\underset{\sim}{\partial}^*} A^{1,1}(G) \to 0 \tag{3.6}$$

generating the exact sequence of cohomology groups

$$0 \to H^0(Q^1(\underset{\sim}{\partial}^*,G)) \xrightarrow{i} H^0(\hat{A}^{1,0}(G)) \xrightarrow{\underset{\sim}{\partial}^*} H^0(A^{1,1}(G)) \to H^1(Q^1(\underset{\sim}{\partial}^*,G)) \to$$

$$\to H^1(\hat{A}^{1,0}(G)) \to H^1(A^{1,1}(G)) \to H^2(Q^1(\underset{\sim}{\partial}^*,G)) \ . \tag{3.7}$$

From

$$H^k(\hat{A}^{1,0}(G)) = H^k(A^{1,1}(G)) = 0 \ , \ k \geq 1 \ ,$$

we obtain the Dolbeault theorem in the form

$$H^1(Q^1(\underset{\sim}{\partial}^*,G)) = H^0(A^{1,1}(G))/\underset{\sim}{\partial}^* H^0(\hat{A}^{1,0}(G)) \ , \tag{3.8}$$

$$H^k(Q^1(\underset{\sim}{\partial}^*,G)) = 0 \ , \ k > 1 \ . \tag{3.9}$$

B. The Carleman-Bers-Vekua operator in spaces of sections of bundles

Let $p_k \in G$ $(k = 1,\dots,n)$ be points of the domain G (remember that G may coincide with M) and α_k be arbitrary integers. The symbol $\gamma = \sum_{k=1}^{n} \alpha_k p_k$ is called a divisor in G. Let $f(p)$ be an analytic function or a covariant in the domain G having no singularities except poles. Then $f(p)$ determines the divisor $(f) = \sum_{k=1}^{n} \alpha_k p_k$ of its poles and zeros p_k, $k = 1,\dots,n$. Here α_k is the order of a pole (if $\alpha_k < 0$) or zero (if $\alpha_k > 0$). Divisors form a group: $\gamma_1 = \sum_k \alpha_k p_k$, $\gamma_2 = \sum_k \alpha'_k p_k$, $\gamma_1 + \gamma_2 = \sum_k (\alpha_k + \alpha'_k)\, p_k$ (if the point p_s does not belong to a divisor, $\alpha_s = 0$). The devisor $\gamma = \Sigma\, \alpha_k p_k \geq 0$ if all $\alpha \geq 0$. The function (covariant) $f(q)$ is a multiple of the divisor γ if $(f) \geq \gamma$. Every divisor $\gamma = \Sigma\, \alpha_k p_k$ corresponds to the number $\deg \gamma = \Sigma\, \alpha_k$ called the degree of the divisor.

As above, let the covering $N_G = \{U_j\ ,\ j \in J_G\}$ of the domain G be some subset of the covering N (it coincides with N if $G = M$) and $\gamma = \Sigma\, \alpha_k p_k$ be some divisor. Any domain U_j of the covering N_G contains some set of the points p_k of γ (this set may be empty) forming the divisor $\gamma_i = \Sigma\, \alpha_k p_k$, $p_k \in U_i$. In the local coordinate $z_i(U_i)$ the points p_k correspond to the points $z_i(p_k)$ and the divisor γ_i corresponds to the divisor $\tilde{\gamma}_i = \Sigma\, \alpha_k z_i(p_k)$ in the z_i-plane. If the function $f(p)$ is a multiple of the divisor γ it is representable in the form

$$f(p) = \gamma_i(p)\, f_{i0}(p) \ , \ \gamma_i(p) = \prod_{p_k \in U_i} (-_i - z_i(p_k))^{\alpha_k} \qquad (3.10)$$

in the domain U_j ; here the function f_{i0} has neither zeros nor poles in U_i . If $p \in U_i \cap U_j$, the value $f_{i0}(p)$ depends on both the point p and the local coordinate. Under the change of the coordinate $z_i(p)$ to $z_j(p)$ we have

$$f_i(p) = f_j(p)\, \gamma_{ij}(p) \ , \ \gamma_{ij}(p) = \frac{\gamma_j(p)}{\gamma_i(p)} \ . \qquad (3.11)$$

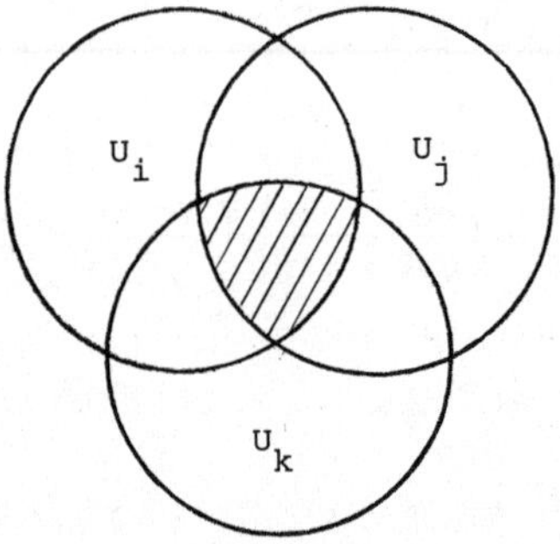

Figure 4

The values $\gamma_{ij}(p)$ are holomorphic functions different from zero defined in the intersections $U_i \cap U_j$. Therefore, they form the 1-cochain $\{\gamma_{ij}\}$ with values in the multiplicative sheaf Ω^* of the germs of holomorphic functions different from zero (see the Appendix).

The analogy between (3.11) and (2.6) is striking. A general approach to the dependence on a local coordinate is realized by the idea of a fibre bundle.

Let $\bar{\gamma} \in H^1(\Omega^*)$ and $\tilde{\gamma} = \{\gamma_{ij}\}$ be a 1-cocycle representing this element. The complex line bundle determined by the cocycle $\tilde{\gamma}$ consists of the space B of the bundle and the continuous mapping $\pi\colon B \to M$ called the projection possessing the following properties. Every set $\pi^{-1}(U_j)$ is homeomorphic to $U \times C$, i.e. in $\pi^{-1}(U_j)$ there are local coordinates $(p_j f_j)$, $\pi(p_j f_j) = p$, $f_j \in C$. For any $i,j \in J$

$$f_i = \gamma_{ij} f_j \quad \text{in} \quad U_i \cap U_j \qquad (3.12)$$

where $\gamma_{ij}(p)$ are holomorphic functions different from zero called transition functions of the bundle B .

Bundles are called equivalent if the cocycles determining the transition functions $\{\gamma_{ij}\}$ and $\{\gamma'_{ij}\}$ belong to one cohomology class $\bar{\gamma}$, i.e. there exists a 0-cochain $\{\gamma_i\} \in Z^0(\Omega^*)$, $\gamma_i(p)$ are holomorphic functions different from zero, such that $\gamma'_{ij} = \gamma_i\gamma_{ij}\gamma_j^{-1}$. Therefore, a bundle is trivial (equivalent to the bundle with transition functions $\gamma_{ij} \equiv 1$) if $\gamma_{ij} = \gamma_i\gamma_j^{-1}$. Note, that if $\{\gamma_{ij}\}$ is a cocycle, in every nonempty intersection $U_i \cap U_j \cap U_k$ the relation

$$\gamma_{ij}\gamma_{jk}\gamma_{ki} = 1$$

is valid.

A section of B over the domain G is a single-valued mapping $f\colon G \to B$ such that $\pi\circ f = 1$. It is clear that the function $f(p)$ depends on both the point p and a local coordinate of p .

The values (3.11) are sections of the bundle B_γ corresponding to the divisor γ and determined by the transition functions $\{\gamma_{ij}\}$ in (3.11) . Covariants of the type (1,0), (0,1), and (1,1) are sections of bundles denoted as $K, \bar{K}$, and $K + \bar{K}$, respectively and determined by the transition functions

$$h_{ij} = \frac{dz_j}{dz_i} , \quad \bar{h}_{ij} = \frac{d\bar{z}_j}{d\bar{z}_i} , \quad h_{ij}\circ\bar{h}_{ij} = \frac{d\bar{z}_j\wedge dz_j}{d\bar{z}_i\wedge dz_i} . \tag{3.13}$$

The sum of bundles B and B' with transition functions $\{\gamma_{ij}\}$ and $\{\gamma'_{ij}\}$ is the bundle B + B' with transition functions $\{\gamma_{ij}\gamma'_{ij}\}$.

Let B be some bundle with transition function $\{\gamma_{ij}\}$ and $a(p)$, $b(p)$ be sections of the bundles $\bar{K}$ and $B - \bar{B} + \bar{K}$ over G . Then one can consider the operator

$$\underset{\sim}{\partial}_B u \equiv \bar{\partial}u + au + b\bar{u} \tag{3.14}$$

transforming sections of the bundle B into sections of the bundle $B + \bar{K}$.

Let $Q(\underset{\sim}{\partial}_B, G)$ be the sheaf of germs of sections of the bundle B which are regular solutions of the equation $\underset{\sim}{\partial}_B u = 0$ in the domain G , $\hat{A}^0(B,G)$ be the sheaf of germs of sections of the bundle B belonging to the classes $\tilde{A}_p(U_i)$ in every coordinate neighborhood U_i and $A^{0,1}(B,G)$ be the sheaf of germs of sections of the bundle $B + \bar{K}$ belonging to the classes $L_p(U_i)$, $p > 2$, in every U_i , i.e.

$$A^{0,1}(B,G) = A^0(B + \bar{K}, G) .$$

We have the exact sequence

$$0 \to Q(\underset{\sim}{\partial}_B, G) \to \hat{A}^0(B,G) \xrightarrow{\underset{\sim}{\partial}_B} A^{0,1}(B,G) \to 0 \tag{3.15}$$

since the Poincaré lemma remains valid in this case. Passing to the sequence of cohomology groups and taking into account that the corresponding sheaves are thin, yields the Dolbeault theorem in the form

$$H^1(Q(\underset{\sim}{\partial}_B, G)) \cong H^0(A^0(B + \bar{K}, G)/\underset{\sim}{\partial}_B H^0(\hat{A}^0(B,G)) ,$$

$$H^k(Q(\underset{\sim}{\partial}_B, G)) = 0 , \ k > 1 . \tag{3.16}$$

In particular, consider the case of the bundle $B = B_\gamma$ determined by the divisor γ.

Let

$$\underset{\sim}{\partial} u = \bar{\partial} u + au + b\bar{u} \tag{3.17}$$

and let us consider solutions of the equation $\underset{\sim}{\partial} u = 0$ which are multiples of the divisor $\underset{\sim}{\partial}$. Then in the domains of the covering $\{U_i\}$ the solutions are represented in the form

$$u(q) = \gamma_i(q)\, u_{i0}(q) . \tag{3.18}$$

Here $\{u_{i0}(q)\}$ are sections of the bundles B_γ satisfying the equation

$$\underset{\sim}{\partial}_{B_\gamma} u_i \equiv \underset{\sim}{\partial}_\gamma u_i \equiv \bar{\partial} u_i + a_i u_i + b_i \bar{u}_i \quad \text{in} \quad U_i ,$$

$$a_i(p) = a(p) , \ b_i(p) = b(p) \frac{\gamma_i(p)}{\overline{\gamma_i(p)}} . \tag{3.19}$$

As it is seen from (3.19) the cochains $a = \{a_i\}$, $b = \{b_i\}$ are sections of the bundles $\bar{K}$ and $\bar{K} + B_\gamma - \bar{B}_\gamma$. Therefore, the sheaf $Q(\underset{\sim}{\partial}_{B_\gamma}, G)$ of germs of generalized analytic sections of the bundle B_γ coincides with the sheaf $Q_{-\gamma}(\underset{\sim}{\partial}, G)$ of germs of generalized analytic functions which are multiples of the divisor γ.

The equation (2.10) keeps within this scheme if $B = K$ is assumed.

§ 4. The Riemann-Roch Theorem

A. Conjugate operators and the Serre duality

1. Let, as above, $a,b \in L_p^{0,1}(M)$, $p > 2$. Consider $\underset{\sim}{\partial}$ as an unbounded operator establishing the mapping

$$\underset{\sim}{\partial}: L_2^0(M) \to L_2^{0,1}(M) .$$

On the product $L_2^{i,j}(M) \times L_2^{1-i,1-j}(M)$ the bilinear form

$$(f,g) = \operatorname{Re} \int_M fg \, d\sigma_t , \tag{4.1}$$

$$f \in L_2^{i,j}(M) , \quad g \in L_2^{1-i,1-j}(M) , \quad i,j = 0,1$$

is invariant. Since the product fg is a covariant of the type (1,1), the integral (4.1) is invariant under changes of local coordinates.

Describing duality by the form (4.1), we obtain the following diagram

$$\begin{array}{ccc} \underset{\sim}{\partial}: L_2^0(M) & \longrightarrow & L_2^{0,1}(M) \\ \downarrow \text{duality} & & \downarrow \text{duality} \\ L_2^{1,1}(M) & \longleftarrow & L_2^{1,0}(M) \; : \underset{\sim}{\partial}^* . \end{array} \tag{4.2}$$

Here the operator

$$\underset{\sim}{\partial}^* v = - \bar{\partial} v + av + \overline{bv} \tag{4.3}$$

is determined on covariants of the type (1,0). The operator $\underset{\sim}{\partial}$ is closed (see Hörmander [a] where the closedness of the operator $\bar{\partial}$ is proved; the operator $\underset{\sim}{\partial}$ differs from $\bar{\partial}$ by an unessential addendum). The diagram involves the relation

$$\operatorname{Ker} \underset{\sim}{\partial}^* \cong \operatorname{Coim} \partial = L_2^{0,1}(M)/\underset{\sim}{\partial} \, L_2^0(M) . \tag{4.4}$$

We have

$$\operatorname{Ker} \underset{\sim}{\partial}^* = \Gamma(Q^1(\underset{\sim}{\partial}^*,M)) = H^0(Q^1(\underset{\sim}{\partial}^*,M) . \tag{4.5}$$

On account of the closedness of the operator $\underset{\sim}{\partial}$, Theorem 1.5, and (3.16) we have

$$L_2^{0,1}(M)/\underset{\sim}{\partial}L_2^0(M) = H^0(A^{0,1}(M))/\underset{\sim}{\partial}H^0(\hat{A}^0(M)) = H^1(Q(\underset{\sim}{\partial},M)) \ . \qquad (4.6)$$

We get the relation

$$H^1(Q(\underset{\sim}{\partial},M) \cong H^0(Q^1(\underset{\sim}{\partial}^*,M)) \qquad (4.7)$$

well-known for the operator $\bar{\partial}$ as the Serre duality theorem (see O. Forster [a], R. Gunning [a]) .

2. The adduced construction can be realized for the operator $\underset{\sim}{\partial}_B$ on sections of the bundle B .

Let $A^{i,j}(B,L_2)$ $(i,j = 0,1)$ be the sheaf of germs of sections of the fibre bundle $B + iK + j\bar{K}$ of the class $L_2(U_j)$ in every domain U_j of the covering N . Here

$$B + iK + j\bar{K} = \begin{cases} B & , \text{if } i = j = 0, \\ B + K & , \text{if } i = 1, j = 0, \\ B + \bar{K} & , \text{if } i = 0, j = 1, \\ B + K + \bar{K} & , \text{if } i = 1, j = 1. \end{cases}$$

Therefore, $A^{i,j}(B,L_2)$ is the sheaf of germs of covariants of the type (i,j)-sections of the space B . The group $\Gamma(A^{i,j}(B,L_2),M) = H^0(A^{i,j}(B,L_2),M)$ is a group of sections of this sheaf over the whole surface M . We consider the operator $\underset{\sim}{\partial}_B$ as the unbounded closed mapping

$$\underset{\sim}{\partial}_B\colon \Gamma(A^0(B,L_2),M) \to \Gamma(A^{0,1}(B,L_2),M) \ .$$

Define the bilinear forms

$$(f,g) = \mathrm{Re}\int_M fg \, d\sigma_p \ ,$$

$$f \in \Gamma(A^{i,j}(B,L_2),M) \ , \ g \in \Gamma(A^{1-i,1-j}(-B,L_2),M), \quad i,j = 0,1 \qquad (4.8)$$

on the products

$$\Gamma(A^{i,j}(B,L_2),M) \times \Gamma(A^{1-i,1-j}(-B,L_2),M) \ , \quad i,j = 0,1 \ .$$

The integral (4.8) is defined correctly since fg is a covariant of the type (1,1), as above.

We have the diagram determining the conjugate operator

$$\begin{array}{ccc} \underset{\sim}{\partial}_B : \Gamma(A^0(B,L_2),M) & \longrightarrow & \Gamma(A^{0,1}(B,L_2),M) \\ \big\downarrow \text{duality} & & \big\downarrow \text{duality} \\ \Gamma(A^{1,1}(-B,L_2),M) & \longleftarrow & \Gamma(A^{1,0}(-B,L_2),M) : \underset{\sim}{\partial}^*_{-B} \end{array} \tag{4.9}$$

which has the form

$$\underset{\sim}{\partial}^*_{-B} v = -\bar{\partial} v + av + \overline{bv} \tag{4.10}$$

and is determined on covariants of the type (1,0) i.e. on sections of the bundle $-B$.

Repeating the considerations adduced above we obtain the Serre duality theorem in the form

$$H^1(Q(\underset{\sim}{\partial}_B, M)) \cong H^0(Q^1(\underset{\sim}{\partial}^*_{-B}, M)) \; . \tag{4.11}$$

B. The Riemann-Roch theorem

Let $\gamma \geq 0$ be some divisor. Consider the exact sequence

$$0 \to Q(\underset{\sim}{\partial}, M) \xrightarrow{i} Q_\gamma(\underset{\sim}{\partial}, M) \xrightarrow{\pi} \gamma \times \mathbf{C} \to 0 \; . \tag{4.12}$$

Evidently, the sheaf of germs of regular solutions of the equation $\underset{\sim}{\partial} u = 0$ is a subset of the sheaf of germs of solutions which are multiples of the divisor $-\gamma$, $Q(\underset{\sim}{\partial}, M) \subset Q_\gamma(\underset{\sim}{\partial}, M)$ and the homomorphism i means an imbedding. The homomorphism π is a projection defined in the following way. In the domain $U_j \in N$, $j \in J$, we have the representation of solutions of the equation $\underset{\sim}{\partial} u = 0$ in the local coordinate z_j

$$u(z_j) = \varphi(z_j) \exp \frac{1}{\pi} \iint_{|t|<1} [a(t) + b(t) \frac{\overline{u(t)}}{u(t)}] \frac{d\sigma_t}{t-z} \; .$$

If the point $p \in U_j$ belongs to the divisor γ with the multiplicity n, then the function $\varphi(z_j)$ has a pole at the point $z_j(p) = \zeta$ of order n with the principal part

$$\text{Pr. part } \varphi(z) = \frac{c_1}{z-\zeta} + \frac{c_2}{(z-\zeta)^2} + \ldots + \frac{c_n}{(z-\zeta)^n} .$$

We associate the pair (p,U_j) with the pair $(p;(c_1,\ldots,c_n))$. The set of such pairs form the sheaf $\gamma \times C$ with the zero-dimensional support γ ; the fibre over every point p is C^n where n is the multiplicity corresponding to p in the divisor γ .

We have (see (A.16))

$$\chi(Q_\gamma(\underset{\sim}{\partial},M)) = \chi(Q(\underset{\sim}{\partial},M)) + \chi(\gamma \times C) . \tag{4.13}$$

In the case $\gamma = \gamma^+ - \gamma^-$, $\gamma^\pm \geq 0$, we consider two exact sequences

$$\begin{gathered} 0 \to Q_{-\gamma^-}(\underset{\sim}{\partial},M) \xrightarrow{i} Q_\gamma(\underset{\sim}{\partial},M) \xrightarrow{\pi} \gamma^+ \times C \to 0 , \\ 0 \to Q_{-\gamma^-}(\underset{\sim}{\partial},M) \xrightarrow{i} Q(\underset{\sim}{\partial},M) \xrightarrow{\pi_1} \gamma^- \times C \to 0 . \end{gathered} \tag{4.14}$$

Here $Q_{-\gamma^-}(\underset{\sim}{\partial},M)$ is the sheaf of germs of solutions of the equation $\underset{\sim}{\partial}u = 0$ which are multipliers of the divisor γ^- . The projection π_1 is constructed in the following manner.

If the function $u(z)$ has a zero in U_j at the point $p \in \gamma^-$ of order m and

$$\varphi(z) = c_0 + c_1(z-\zeta) + \ldots + c_{m-1}(z-\zeta)^{m-1} , \quad \zeta = z_j(p)$$

then the pair (p,U_j) is associated with the pair $(p,(c_0,\ldots,c_{m-1}))$.

From (4.14) if follows that

$$\chi(Q_\gamma(\underset{\sim}{\partial},M)) = \chi(Q_{-\gamma^-}(\underset{\sim}{\partial},M)) + \chi(\gamma^+ \times C) ,$$

$$\chi(Q(\underset{\sim}{\partial},M)) = \chi(Q_{-\gamma^-}(\underset{\sim}{\partial},M)) + \chi(\gamma^- \times C) .$$

Then

$$\chi(Q_\gamma(\underset{\sim}{\partial},M)) = \chi(Q(\underset{\sim}{\partial},M)) + \chi(\gamma^+ \times C) - \chi(\gamma^- \times C) . \tag{4.15}$$

Since the supports of the sheaves $\gamma^\pm \times C$ are zero-dimensional, all their cohomology groups, except the zero one, are equal to zero and hence

$$\chi(\gamma^{\pm} \times C) = \deg \gamma^{\pm} .$$

Therefore,

$$\chi(Q_{\gamma}(\underset{\sim}{\partial},M)) = \chi(Q(\underset{\sim}{\partial},M)) + \deg \gamma . \tag{4.16}$$

As it was established above (see (3.5) and (3.16)), the second and higher cohomology groups of the sheaves $Q(\underset{\sim}{\partial},M)$ and $Q_{\gamma}(\underset{\sim}{\partial},M) = Q(-B_{\gamma},M)$ are equal to zero.

We obtain the relation

$$\dim H^0(Q_{\gamma}(\underset{\sim}{\partial},M)) - \dim H^0(Q^1_{-\gamma}(\underset{\sim}{\partial}^*,M)) =$$

$$= \deg \gamma + \dim H^0(Q(\underset{\sim}{\partial},M)) - \dim H^0(Q^1(\underset{\sim}{\partial}^*,M)) . \tag{4.17}$$

Here we used the definition of the Eulerian characteristic (A.15) and the Serre duality theorem for the bundle $B = B_{\gamma}$ and for trivial B (4.11) and (4.7) and the relation

$$Q^1(\underset{\sim}{\partial}^*_{-B_{\gamma}},M) = Q^1_{-\gamma}(\underset{\sim}{\partial}^*,M) .$$

As it can be seen from equation (4.17), the left hand side of this relation is equal to the index of the operator $\underset{\sim}{\partial}_{\gamma}$ in the space $H^0(Q(\underset{\sim}{\partial}_{B_{\gamma}},M))$ and is determined by the degree of the divisor γ. Below we shall calculate the right hand side of (4.17) and shall show that

$$\dim H^0(Q(\underset{\sim}{\partial},M)) - \dim H^0(Q^1(\underset{\sim}{\partial}^*,M) = 2 - 2g \tag{4.18}$$

where g is the genus of the surface M.

Equation (4.17) is the Riemann-Roch theorem. For $a = b = 0$ it is degenerated to the well-known theorem of algebraic function theory.

C. Inhomogeneous equations

In conclusion of this section, we consider the inhomogeneous equation $\underset{\sim}{\partial}u = F$ on the closed Riemann surface M.

Returing to the diagram (4.2) we formulate the following statement.

<u>Theorem 4.1</u>. In order that the equation $\underset{\sim}{\partial} u = F$, $F \in L_p^{0,1}(M)$, $p > 2$ is solvable it is necessary and sufficient that

$$\mathrm{Re} \int_M F(q)\, v(q)\, d\sigma_q = 0 \tag{4.19}$$

where $v(q)$ is an arbitrary regular solution of the equation $\underset{\sim}{\partial}^* v = 0$.

Note that the integral (4.19) is defined correctly since $v(q)$ is a covariant of the type (1,0) and, consequently, the product $F(q)\, v(q)$ is a covariant of the type (1,1).

Evidently, equation (4.19) is a consequence of the closedness of the operator $\underset{\sim}{\partial}$. The following generalization of Theorem 4.1 is valid, too.

<u>Theorem 4.1'</u>. In order that the equation $\underset{\sim}{\partial}_B u = F$, $F \in \Gamma(A^{0,1}(B,L_2),M)$, is solvable it is necessary and sufficient that

$$\mathrm{Re} \int_M F(q)\, v(q)\, d\sigma_q = 0 \tag{4.19'}$$

where $v(q)$ is an arbitrary regular solution of the equation $\underset{\sim}{\partial}{}_B^* v = 0$.

CHAPTER 2

LINEAR INTEGRAL EQUATIONS CONNECTED WITH GENERALIZED ANALYTIC FUNCTIONS

§ 5. Integral representation kernels

A. Abelian differentials

The integral representations (1.28) and (1.31) playing a principial role for generalized analytic functions are based on the use of the fundamental solution $(z-t)^{-1}$ of the Cauchy-Riemann equation $\bar{\partial}u = 0$ in the plane.

To obtain analogous representations on a Riemann surface we need an analogous kernel. Such kernels were constructed by several authors (H. Titz [a], H. Behnke, K. Stein [a]). We follow the later paper by S.Ja. Gusman, Yu. L. Rodin [a].

First of all, we list the Abelian differentials and integrals which will be used below (see O. Forster [a], R. Gunning [a], G. Springer [a]). The term "an Abelian differential" denotes an invariant differential form $\omega = \varphi(z(p))\,dz(p)$ ($z = z(p)$ is a local coordinate of the point p) where $\varphi(z(p))$ is an analytic covariant of the type (1,0) without essential singularities.

If $\varphi(z(p))$ has no singularities on M, the corresponding differential is called an Abelian differential of the first kind. By the famous Riemann theorem the complex dimension of the linear space of Abelian differentials of the first kind is equal to g. We shall use the basis of this space dw_j $(j = 1,\ldots,g)$ normalized by the period relations along the cycles K_{2j-1} $(j = 1,\ldots,g)$ of the homology basis

$$\int_{K_{2j-1}} dw_k = \delta_{jk}\ , \quad j,k = 1,\ldots,g\ . \tag{5.1}$$

We consider also the real basis of this space $d\theta_j$ $(j = 1,\ldots,2g)$ with periods

$$\operatorname{Im}\int_{K_{2j-1}} d\theta_{2k} = -\operatorname{Im}\int_{K_{2j}} d\theta_{2k-1} = \delta_{kl}\ ,$$

$$\operatorname{Im} \int_{K_{2_j}} d\theta_{2k} = \operatorname{Im} \int_{K_{2j-1}} d\theta_{2k-1} = 0 , \tag{5.2}$$

$k,j = 1,\dots,g$.

The residue of an Abelian differential ω at the point $p_0 \in M$ is the value

$$\operatorname{Res}_{p_0} \omega = \frac{1}{2\pi i} \int_{l_{p_0}} \omega \tag{5.3}$$

where l_{p_0} is a small closed contour around the point p_0 . If all residues of an Abelian differential are equal to zero, it is called an Abelian differential of the second kind. We use normalized Abelian differentials of the second kind $dt^n_{q,z}(s)$, $dT^n_{q,z}(s)$ having a single pole of order $n+1$ at the point $q \in M$ $(n \geq 1)$ with the principal parts expressed in the fixed local coordinate $z(q)$ as

$$\text{Pr. part } dt^n_{q,z} = \text{Pr. part } dT^n_{q,z} = - \frac{n\, dz(s)}{[z(s)-z(q)]^{n+1}} \tag{5.4}$$

and periods

$$\begin{gathered} \int_{K_{2j-1}} dt^n_{q,z}(s) = 0 , \quad j = 1,\dots,g , \\ \operatorname{Re} \int_{K_j} dT^n_{q,z}(s) = 0 , \quad j = 1,\dots,2g . \end{gathered} \tag{5.5}$$

As a rule in the case $n = 1$ the symbol n will be omitted. The subscript z often is omitted, too.

Let q_0 and q be two arbitrary different points of the surface. The normalized Abelian differentials of the third kind $d\omega_{q_0q}(s)$ and $d\Omega_{q_0q}(s)$ have poles of first order with residues ∓ 1 at the points q_0 and q , respectively and periods

$$\int_{K_{2j-1}} d\omega_{q_0q}(s) = 0 , \quad j = 1,\dots,g , \tag{5.6}$$

$$\operatorname{Re}\int_{K_j} d\Omega_{q_0 q}(s) = 0\ , \quad j = 1,\dots,2g\ . \tag{5.6}$$

Now several known relations connecting Abelian differentials (see, for example, G. Springer [a]) are listed.

$$\int_{K_{2j}} dw_k = \int_{K_{2k}} dw_j\ , \quad k,j = 1,\dots,g\ , \tag{5.1'}$$

$$\int_{K_j} d\Omega_{q_1 q_2}(s) = 2\pi i \operatorname{Im}\int_{q_1}^{q_2} d\theta_j\ , \quad j = 1,\dots,2g\ , \tag{5.7}$$

$$\int_{K_{2j}} d\omega_{q_1 q_2}(s) = 2\pi i \int_{q_1}^{q_2} dw_j\ , \quad j = 1,\dots,g\ , \tag{5.8}$$

$$\int_{K_{2j}} dt^n_{q,z}(s) = -\frac{2\pi i}{(n-1)!}\frac{d^n\theta_j(q)}{d[z(q)]^n}\ , \quad j = 1,\dots,g\ , \tag{5.9}$$

$$\int_{K_j} dT^n_{q,z}(s) = -\frac{2\pi i}{(n-1)!}\operatorname{Im}\frac{d^n\theta_j(q)}{d[z(q)]^n}\ , \quad j = 1,\dots,2g\ , \tag{5.10}$$

$$\int_{q_1}^{q_2} dt^n_{q,z}(s) = -\frac{1}{(n-1)!}\frac{d^n\omega_{q_1 q_2}(q)}{d[z(q)]^n}\ , \tag{5.11}$$

$$\omega_{s_0 s}(q) - \omega_{s_0 s}(q_0) = \omega_{q_0 q}(s) - \omega_{q_0 q}(s_0)\ , \tag{5.12}$$

$$\Omega_{s_0 s}(q) - \Omega_{s_0 s}(q_0) = \Omega_{q_0 q}(s) - \Omega_{q_0 q}(s_0) -$$

$$- 2\pi i \sum_{j=1}^{g}\left[\operatorname{Im}\int_{q_0}^{q} d\theta_{2j-1}\operatorname{Im}\int_{s_0}^{s} d\theta_{2j} - \operatorname{Im}\int_{q_0}^{q} d\theta_{2j}\operatorname{Im}\int_{s_0}^{s} d\theta_{2j-1}\right]\ . \tag{5.13}$$

Note, that if df is an Abelian differential, $f(p)$ is a multivalued function called an Abelian integral.

In conclusion, we adduce the Riemann-Roch theorem for analytic functions.

Let γ be some divisor on M, $\gamma = \Sigma \alpha_k p_k$. Consider the space $L(\gamma)$ of functions which are multiples of the divisor $-\gamma$ and the space $H(\gamma)$ of Abelian differentials which are multiples of the divisor γ. Then

$$\dim L(\gamma) - \dim H(\gamma) = \deg \gamma - g + 1 \tag{5.14}$$

where the dimensions are complex.

B. Multi-valued kernels

The Cauchy kernel $(t-z)^{-1}\,dt$ on the plane possesses the following obvious properties.

- It is an Abelian differential of the third kind with respect to t with poles $t = z$, ∞ with residues ± 1, respectively.
- It is an analytic function with respect to z with a pole of first order at $z = t$ and a zero of the first order at $z = \infty$.

Assume

$$m(s,q) = \partial_s\,[\omega_{s_0 s}(q) - \omega_{s_0 s}(q_0)]\,, \tag{5.15}$$

$$M(s,q) = \partial_s\,[\Omega_{s_0 s}(q) - \Omega_{s_0 s}(q_0)]\,. \tag{5.16}$$

Here

$$\partial_s = \frac{1}{2}\left(\frac{\partial}{\partial x} - i\,\frac{\partial}{\partial y}\right), \qquad z = x + iy = z(s)\,.$$

The points s_0 and q_0 are supposed to be fixed in the sequel.

Both the forms (5.15) and (5.16) can be used as Cauchy kernels. They will be studied now.

It is clear that both these forms have poles of first order at the point $q = s$ and are Abelian integrals of the second kind with respect to q (see (5.11) for $n = 1$). It is more hard to ascertain the properties with respect to the first variable. For the form $m(s,q)$ one can get them from equation (5.12):

$$m(s,q)\ dz(s) = d\omega_{q_0 q}(s) .$$

We infer that the form $m(s,q)$ is an Abelian covariant of the third kind with the poles $s = q_0$, q with residues ∓ 1, respectively. At last, from (5.14) it follows that

$$m(s,q_0) \equiv 0 . \tag{5.17}$$

For the form $M(s,q)$ we obtain from (5.13)

$$M(s,q)\ dz(s) = d\Omega_{q_0 q}(s) - \\ - \pi \sum_{j=1}^{g} [(\text{Im} \int_{q_0}^{q} d\theta_{2j-1}) d\theta_{2j}(s) - (\text{Im} \int_{q_0}^{q} d\theta_{2j}) d\theta_{2j-1}(s)] . \tag{5.18}$$

Here we used the relation

$$\partial(\text{Im } f(z)) = \partial\ (\frac{f(z)-\overline{f(z)}}{2i}) = \frac{1}{2i} f'(z)$$

which is valid for analytic functions.

From (5.18) it follows that $M(s,q)$ is also an Abelian covariant of the third kind with the residues ∓ 1 at the points $s = q_0, q$.

Periods of the kernels (5.15) and (5.16) are equal by (5.7), (5.8)

$$l_j(s) = \int_{K_{2j}} d_q\ m(s,q) = \partial_s\ 2\pi i \int_{s_0}^{s} dw_j = 2\pi i\ w_j'(s) , \\ \int_{K_{2j-1}} d_q\ m(s,q) = 0 , j = 1,\dots,g . \tag{5.19}$$

$$L_j(s) = \int_{K_j} d_q\ M(s,q) = \partial_s\ 2\pi i\ \text{Im} \int_{s_0}^{s} d\theta_j = \\ = \pi\theta_j'(s) , \quad j = 1,\dots,2g . \tag{5.20}$$

From (5.16) we have

$$M(s,q_0) = 0 \,. \tag{5.17'}$$

In conclusion we consider the Abelian integral (5.13)

$$\tilde{M}(s,q) = \Omega_{s_0 s}(q) - \Omega_{s_0 s}(q_0) = \Omega_{q_0 q}(s) - \Omega_{q_0 q}(s_0) -$$

$$- 2\pi i \sum_{j=1}^{g} \{\mathrm{Im} \int_{q_0}^{q} d\theta_{2j-1}\, \mathrm{Im} \int_{s_0}^{s} d\theta_{2j} - \mathrm{Im} \int_{q_0}^{q} d\theta_{2j}\, \mathrm{Im} \int_{s_0}^{s} d\theta_{2j-1}\} \tag{5.13}$$

as a function of the variable s.

We calculate the increment of the value of $\tilde{M}(s,q)$ when the point s moves around the cycle K_l in the positive direction. By equations (5.2), (5.7) we have

$$\Delta_{K_l} \tilde{M}(s,q) = \int_{K_l} d\Omega_{q_0 q}(s) - 2\pi i \sum_{j=1}^{g} [\mathrm{Im} \int_{q_0}^{q} d\theta_{2j-1}\, \mathrm{Im} \int_{K_l} d\theta_{2j}$$

$$- \mathrm{Im} \int_{q_0}^{q} d\theta_{2j}\, \mathrm{Im} \int_{K_l} d\theta_{2j-1}] = 2\pi i\, \mathrm{Im} \int_{q_0}^{q} d\theta_l - 2\pi i \int_{q_0}^{q} \mathrm{Im}\, d\theta_l = 0 \,,$$

$$l = 1, \ldots, 2g \,. \tag{5.20'}$$

Therefore, the value $d\tilde{M}(s,q)$ has zero periods while the kernel (5.16) has periods with respect to the variable s.

C. The single-valued Cauchy kernel

If the use of multi-valued functions is undesirable it is convenient to have a single-valued Cauchy kernel (S.Ja. Gusman, Yu. L. Rodin [a]).

Let $\delta = \sum_k \alpha_k q_k \geq 0$ be some divisor such that $\dim H(\delta) = 0$ (see § 5A). By the Riemann-Roch theorem (5.14)

$$\dim H(\delta) = \dim L(\delta) + g - 1 - \deg \delta \,.$$

Hence from $\dim H(\delta) = 0$ it follows that $\deg \delta \geq g$. Consider the complex normalized Abelian integral of the second kind which is a multiple of the divisor $-\delta$

$$t_\delta(q) = \sum_k \sum_{l=1}^{\alpha_k} c_{kl}\, t^l_{q_k}(q) \tag{5.21}$$

(we omit the symbol z in (5.4), (5.5)).
The coefficients c_{kl} we reserve to be indeterminate for the present. Periods of this integral along the cycles K_{2j} are equal to

$$l_{\delta_j} = \int_{K_{2j}} dt_\delta = \sum_k \sum_{\ell=1}^{\alpha_k} c_{kl} \int_{K_{2j}} dt^l_{q_k}\,, \qquad j = 1,\dots,g\,.$$

The system

$$\sum_k \sum_{l=1}^{\alpha_k} c_{kl} \int_{K_{2j}} dt^l_{q_k} = 0\,, \qquad j = 1,\dots,g\,,$$

has $\dim L(\delta) - 1$ solutions since every solution corresponds to a single-valued function (5.21) which is a multiple of the divisor $-\delta$ and all functions of the space $L(\delta)$ except 1 are representable in the form (5.21). By the Riemann-Roch theorem we have

$$\dim L(\delta) = \deg \delta - g + 1$$

since $H(\delta) = 0$. Therefore, the rank r of the matrix

$$\Big[\int_{K_{2j}} dt^l_{q_k}\Big]\,, \; j = 1,\dots,g\,;\; l = 1,\dots,\alpha_k\,;\; k = 1,2,\dots$$

is equal to $r = \deg\delta - (\deg\delta - g + 1 - 1) = g$.
Hence the inhomogeneous system

$$\sum_k \sum_{i=1}^{\alpha_k} c_{ki} \int_{K_{2j}} dt^i_{q_k} = l_j\,, \qquad j = 1,\dots,g \tag{5.22}$$

is solvable for arbitrary right hand side.

Let $l_j = l_j(s)$ be the periods (5.19) of the kernel $m(s,q)$. From equation (5.22) we then obtain coefficients $c_{ki} = c_{ki}(s)$. Consider the form

$$m_\delta(s,q)\, dz(s) = m(s,q)\, dz(s) - \sum_k \sum_{i=1}^{\alpha_k} c_{ki}(s)\, t^i_{q_k,z_k}(q)\, dz(s)\,. \tag{5.23}$$

It has poles $s = q$ and $s = q_0$ and additional new poles determined by the divisor δ. This divisor is called characteristic for the kernel. It is clear that the kernel (5.23) is single-valued with respect to both variables.

In the case $\deg \delta = g$ we denote the divisor δ as minimal. For such a divisor $\dim L(\delta) = g - g + 1 = 1$. It means that there exists no analytic function different from a constant which is a multiple of the divisor $-\delta$.

Consider until further notice the case of a divisor δ without multiple points, $\delta = \sum_{k=1}^{g} q_k$. Then equation (5.22) has the form

$$\sum_{k=1}^{g} c_k \frac{dw_j(q_k)}{dz_k(q)} = \frac{dw_j(s)}{dz(s)}, \quad j = 1,\ldots,g$$

where $z_k(q)$ are fixed coordinates of the points q_k and the local coordinate $z(q)$ in the right hand side is determined by the chosen integral $t_{q,z}(q)$. It is clear that solutions of this system are Abelian covariants of the first kind

$$c_k(s)dz(s) = dZ_k(s), \quad k = 1,\ldots,g$$

determined by the conditions

$$\frac{dZ_j(q_k)}{dz_k(q)} = \delta_{kj}, \quad k,j = 1,\ldots,g. \tag{5.24}$$

The coordinates $z_k(q)$ are determined by the integrals $t_{q_k,z_k}(q)$ used in (5.23). In this case we have

$$m_\delta(s,q) = m(s,q) - \sum_{k=1}^{g} t_{q_k}(q) Z_k'(s). \tag{5.25}$$

The properties of this kernel are illustrated by figure 5.

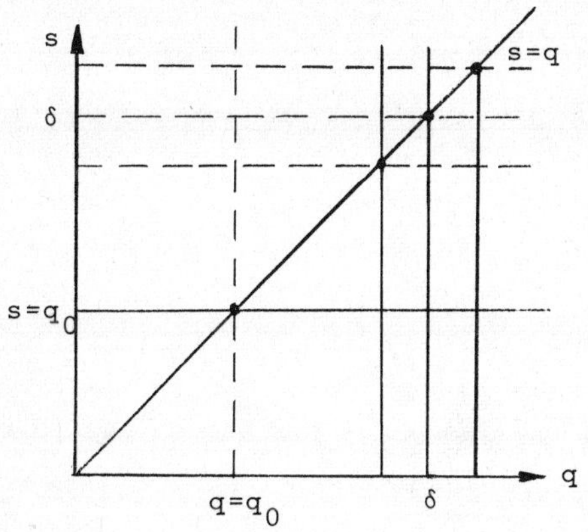

Figure 5

Here solid lines mean poles and dotted lines correspond to zeros of the kernel. The diagonal line corresponds to the pole $s = q$ and solid vertical lines mean the poles $q = q_k$. The horizontal dotted lines conform with zeros $m_\delta(q_k,q) = 0$. This follows from (5.24) and the relation

$$m(q_k,q) = - t_{q_k}(q) \ , \quad k = 1,\ldots,g \ . \tag{5.26}$$

At last, the vertical dotted line $m_\delta(q,q_0) = 0$ is provided by the corresponding choice of branches of all integrals in (5.25)

$$t_{q_k}(q_0) = 0 \ , \quad k = 1,\ldots,g \ . \tag{5.27}$$

At the solid nodes the kernel is regular and is equal to zero. At the points (q_k,q_k) $(k = 1,\ldots,g)$ the kernel is equal to zero by (5.25) and (5.26), $m(q_k,q_0) = 0$ $(k = 0,\ldots,g)$ by (5.27).

In conclusion note that in the fixed coordinates $z_k(q_k)$ the principal parts of the poles $q = q_k$ are equal to $- Z_k'(s)$ by equations (5.4), (5.25).

§ 6. Integral equations

A. The integral operator P

In this section we use the multi-valued Cauchy kernel $m(s,q)$ (5.15) to study multi-valued solutions of the Carleman-Bers-Vekua system. Slit the surface M along cycles $K_j (j = 1,\dots,2g)$ of the homology basis and transform M by this cutting into the 4g-sided polygon $\hat{M}$ (figure 3). It is clear that in this plygon one can choose a single-valued branch of every multi-valued function on M. In particular we can fix the value of the function at some point q_0. The chosen branch has jumps on the banks of the cuts (see figure 6).

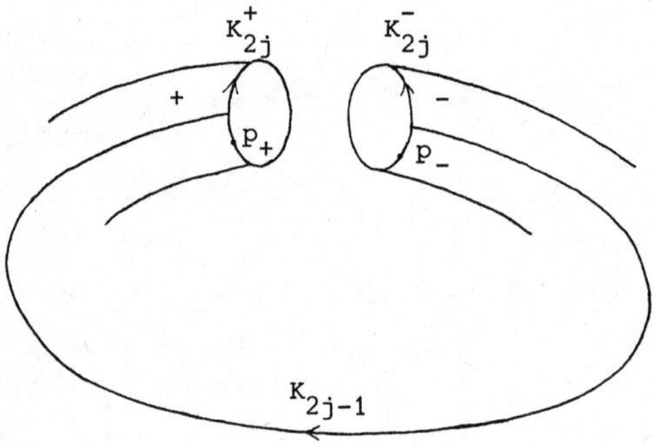

Figure 6

The value of the jump is equal to the period of the integral along the adjoint cycle

$$\Delta_{2j} f = f(p_+) - f(p_-) = \int_{K_{2j-1}} df \,,$$

$$\Delta_{2j-1} f = f(p_+) - f(p_-) = - \int_{K_{2j}} df \,. \tag{6.1}$$

Let $a(s), b(s) \in L_p^{0,1}(M)$, $p > 2$, be coefficients of the Carleman-Bers-Vekua equation. Analogously to (1.23), (1.27) introduce the operator

$$Pu = -\frac{1}{\pi}\iint\limits_{\hat{M}} [a(s)\,u(s) + b(s)\,\overline{u(s)}]\, m(s,q)\, d\sigma_s \,. \tag{6.2}$$

The integral (6.2) is defined correctly since the value under the integral is a covariant of the type (1,1). The branch of the multi-valued function $m(s,q)$ is fixed by the condition

$$m(s,q_0) = 0 \,. \tag{6.3}$$

The choice of the branch of the multi-valued function $u(s)$ will be stipulated if it is necessary.

Theorem 6.1. The function $h(q) = Pu(q)$ determined by (6.2) is a multi-valued function on M having zero periods along the cycles K_{2j-1} $(j = 1,\ldots,g)$. The periods of $h(q)$ along the cycles K_{2j} are equal to

$$h_j = -2i \iint\limits_{\hat{M}} [a(s)\,u(s) + b(s)\,\overline{u(s)}]\, w_j'(s)\, d\sigma_s \,,\ j = 1,\ldots,g \,. \tag{6.4}$$

The branch of $h(q)$ on $\hat{M}$ corresponding to the chosen branch of the kernel (6.3) is characterized by the condition $h(q_0) = 0$.

Directly from (5.19) the next statement follows.

Theorem 6.2. Let $a(s), b(s) \in L_p^{0,1}(M)$, $p > 2$. Then the operator (6.2) is completely continuous in the space $C(\hat{M})$ and maps this space into the space $C_\alpha(\hat{M})$, $\alpha = \frac{p-2}{p}$, and

$$\| Pu \|_{C_\alpha(\hat{M})} \le M_p \{\| a \|_{L_p(M)} + \| b \|_{L_p(M)}\} \| u \|_{C(\hat{M})} \,.$$

This operator is completely continuous also in the spaces $L_q(\hat{M})$, $\frac{1}{2} \le \frac{1}{p} + \frac{1}{q} \le 1$ and maps it into the spaces $L_\gamma(\hat{M})$, $\frac{1}{p} + \frac{1}{q} - \frac{1}{2} < \frac{1}{\gamma} < 1$. In particular, it is valid for $\gamma = q$.

Note. If an integer n satisfies the condition

$$n - 1 \le \frac{2p}{p-2}\left(\frac{1}{p} + \frac{1}{q} - \frac{1}{2}\right) < n \,,$$

then

$$\| P^{n+1}u \|_{C_\beta(\hat{M})} \le M'_{p,q,\alpha} \{ \| a \|_{L_p(M)} + \| b \|_{L_p(M)} \} \| u \|_{L_p(\hat{M})} ,$$

$$\| P^k u \|^{\alpha}_{L_{\gamma_k}(\hat{M})} \le M_{p,q,\alpha} \{ \| a \|_{L_p(M)} + \| b \|_{L_p(M)} \} \| u \|_{L_q(\hat{M})} ,$$

$k = 1,\dots,n$,

where

$$\frac{1}{\gamma_k} = \frac{1}{q} + \frac{k}{p} - \frac{k}{2} + k\alpha , \quad k = 1,\dots,n ,$$

$$\beta = 1 - 2\left(\frac{1}{p} + \frac{n+1}{p} - \frac{n}{2} + n\alpha\right) ,$$

$$0 < \alpha < \frac{p-2}{2p} - \frac{1}{n}\left(\frac{1}{p} + \frac{1}{q} - \frac{1}{2}\right) .$$

Let $\mathcal{M} = \{U_i, i \in I\}$ be a triangulation of the surface M. We assume that the cycles K_j $(j = 1,\dots,g)$ are formed by sides of triangles of $\mathcal{M}$. Then

$$Pu = \sum_{i \in I} P_i u ,$$

$$h_i(q) = P_i u = -\frac{1}{\pi} \iint_{U_i} [a(s)u(s) + b(s)\overline{u(s)}] m(s,q)\, d\sigma_s . \qquad (6.5)$$

For the domain U_i and the point q_0, $q_0 \in U_i$, we have

$$h_i(z) = -\frac{1}{\pi} \iint_{z(U_i)} \frac{a(t)u(t) + b(t)\overline{u(t)}}{t-z}\, d\sigma_t +$$

$$+ \frac{1}{\pi} \iint_{z(U_i)} \frac{a(t)u(t) + b(t)\overline{u(t)}}{t-z_0}\, d\sigma_t +$$

$$+ \frac{1}{\pi} \iint_{z(U_i)} [a(t)u(t) + b(t)\overline{u(t)}] m_0(t,z)\, d\sigma_t , \; z = z(q) , \; z_0 = z(q_0) ,$$

where $m_0(t,z)$ is an analytic function. Each function $h_i(q)$ satisfies the inequality

$$\| h_i(q) \|_{C_\alpha(U_i)} \le M_{p,i}\{\| a \|_{L_p(U_i)} + \| b \|_{L_p(U_i)}\} \| u \|_{C(U_i)}$$

and maps the space $L_q(U_i)$ into $L_\gamma(U_i)$ (Theorems 1.2 and 1.5) completely continuously. The function $h_i(q)$ is analytic in the domain $\hat{M} - U_i$ since for $q \in \hat{M} - U_i$ the kernel of the operator P_i is analytic.

The inegualities

$$|h_i(q)| \le M_i'\{\| a \|_{L_p(U_i)} + \| b \|_{L_p(U_i)}\} \| u \|_{X(U_i)}$$

are valid; here $X(U_i)$ is the space $C(U_i)$ or $L_q(U_i)$. These inequalities entail the theorem's statement.

On the product $L_q^0(M) \times L_{q'}^{1,0}(M)$, $\frac{1}{2} \le \frac{1}{q} + \frac{1}{p} \le 1$, $\frac{1}{p} + \frac{1}{q} + \frac{1}{q'} = 1$, we consider the bilinear form

$$(u,v) = \frac{1}{\pi} \operatorname{Re} \iint_{\hat{M}} (au + b\bar{u})v \, d\sigma_s = \frac{1}{\pi} \operatorname{Re} \iint_{\hat{M}} (av + \overline{bv})u \, d\sigma_s . \tag{6.6}$$

We obtain the expression for the conjugate operator

$$P^*v = -\frac{1}{\pi} \iint_{\hat{M}} [a(s)v(s) + \overline{b(s)}\,\overline{v(s)}]m(q,s) \, d\sigma_s \tag{6.7}$$

which is compact in the space $L_{q'}^{1,0}(M)$, $\frac{1}{q'} + (\frac{1}{p} + \frac{1}{q}) = 1$, $q' \ge 2$.

<u>Note</u>. The consideration presented above is not strict since the bilinear form (6.6) can in general degenerate on some pairs (u,v) . Instead of (6.6), one can use the bilinear form

$$(u,w)_1 = \operatorname{Re} \iint_{\hat{M}} u(s)\overline{w(s)} \, d\rho(s) \tag{6.8}$$

where $u \in L_q^0(M)$, $w \in L_{\tilde{q}}^0(M)$, $\frac{1}{2} \le \frac{1}{p} + \frac{1}{q} \le 1$, $\frac{1}{q} + \frac{1}{\tilde{q}} = 1$, and $d\rho(s)$ is a piecewise continuous invariant measure different from zero everywhere on M (such a measure may be chosen in every domain U_i , $i \in I$ of M). It is easy to see that the equation conjugate to $u + Pu = 0$ relatively to the form (6.8) coincides with the equation

$v + P^*v = 0$.

We have the operator $\widetilde{P}^*$ conjugate to P relatively to the bilinear form (6.8)

$$\widetilde{P}^*w = -\frac{a(r)d\sigma_r}{\pi d\rho(r)} \iint_{\hat{M}} w(s)m(r,s)d\rho(s) -$$

$$-\frac{\overline{b(r)}d\sigma_r}{\pi d\rho(r)} \iint_{\hat{M}} \overline{w(s)}\,\overline{m(r,s)}d\rho(s) \ , \quad r \in M \ . \tag{6.9}$$

Consider the equation

$$w + \widetilde{P}^*w = 0 \ .$$

Assume

$$h(s) = \frac{1}{\pi} \iint_{\hat{M}} w(s)m(r,s)d\rho(s) \ . \tag{6.10}$$

We obtain the relation

$$w(s) = [a(s)h(s) + \overline{b(s)}\,\overline{h(s)}]\frac{d\sigma_s}{d\rho(s)} \ . \tag{6.10'}$$

It means that $h \in L^{1,0}_{q'}(M)$, $\frac{1}{q'} + \frac{1}{q} = \frac{1}{\tilde{q}}$, i.e. $\frac{1}{q} + \frac{1}{q'} + \frac{1}{p} = 1$. The measure $d\rho(s)$ is representable in the form $d\rho(s) = |\rho'(s)|^2 \, d\sigma_s$ where $\rho'(s)$ is a covariant of the type (1.1) and, consequently, the relation (6.10') is valid for an arbitrary choice of a local coordinate. Substituting (6.10') into (6.10) we obtain the equation $h + P^*h = 0$. Therefore, the following result is valid.

<u>Theorem 6.3</u>. The equation $u + Pu = 0$ has a finite number g_0 of solutions in the space $L^0_q(M)$, $\frac{1}{2} \le \frac{1}{q} + \frac{1}{p} \le 1$. In order that the inhomogeneous equation $u + Pu = f$ is solvable it is necessary and sufficient that

$$(f,v_j) = 0 \ , \quad j = 1,\dots,g_0 \tag{6.11}$$

where v_j $(j = 1,\dots,g_0)$ is a complete system of solutions of the equation $v + P^*v = 0$.

Note, that the number g_0 is independent on the point q of M since every solution of the equation $u + Pu = 0$ belongs to the space $C_\alpha(\hat{M})$, $0 < \alpha < \frac{p-2}{2p} - \frac{1}{n}(\frac{1}{p} + \frac{1}{q} - \frac{1}{2})$. The same statement holds for equation (1.33).

Consider the covariant $w(s)$ of the type (1,0) determined by the relation

$$w(s) = P^*v = -\frac{1}{\pi}\iint_{\hat{M}} [a(r)v(r) + \overline{b(r)}\,\overline{v(r)}]m(s,r)\,d\sigma_r .$$

<u>Theorem 6.4</u>. The covariant $w(s)$ has a pole of first order at the point $s = q_0$ with the principal part

$$\text{Pr. part}_{s=q_0}\, w(s) = \frac{1}{\pi}\iint_{\hat{M}} [a(r)v(r) + \overline{b(r)}\,\overline{v(r)}]\,d\sigma_r =$$

$$= (1,v) - i(i,v) . \tag{6.12}$$

Here one has used the local coordinate fixed in the integrals (5.4)-(5.5).

The operator P^* completely continuously maps the space $C^{1,0}(\hat{M})$ into the space $C_\alpha^{1,0}(\hat{M}_0)$, $\alpha = \frac{p-2}{2}$ and

$$\| P^*w \|_{C_\alpha} \le M_p^* \{\| a \|_{L_p(M)} + \| b \|_{L_p(M)}\} \| w \|_{C(\hat{M}_0)} .$$

Here $\hat{M}_0$ is the surface $\hat{M}$ without a neighborhood of the point q_0 . The operator P^* is completely continuous in the spaces $L_q^{1,0}(M)$, $\frac{1}{2} \le \frac{1}{p} + \frac{1}{q} \le 1$, too and maps them into the spaces $L_\gamma^{1,0}(M)$, $1 < \gamma < 2$. In particular, this holds for $\gamma = q$ with $1 < q < 2$.

For the solvability of the equation

$$v + P^*v = f$$

in the space $L_q^{1,0}(M)$, $1 < q < 2$, it is necessary and sufficient that

$$(f,u_j) = 0 , \quad j = 1,\ldots,g_0 , \tag{6.13}$$

where $u_j (j = 1,\ldots,g_0)$ is a complete system of solutions of the equation $u + Pu = 0$.

This statement is proved in the same manner as Theorem 6.2. It is necessary to take into account that the covariant P^*f has a pole of first order and hence belongs to the space $L_\gamma^{1,0}(M)$ only for $1 < \gamma < 2$.

Note. The covariant $w = P^*v$ can possess a single pole of first order since for $a,b \neq 0$ the differential $w\,dz(s)$ is not closed.

B. Representations of solutions

We shall consider multi-valued solutions of the equation

$$\underset{\sim}{\partial} u = \bar{\partial} u + au + b\bar{u} = 0 . \tag{6.14}$$

A multi-valued solution of (6.14) is a multi-valued function a branch of which chosen on $\hat{M}$ satisfies the equation (6.14). Note that the other branches of this function in general do not satisfy (6.14) on $\hat{M}$. We shall consider integral representations for the chosen branch.

As it follows from the representation (1.28), every regular solution on M of equation (6.14) is representable in the form

$$u(s) - \frac{1}{\pi} \iint_{\hat{M}} [a(r)u(r) + b(r)\overline{u(r)}]m(r,s)\, d\sigma_r = F(s) \tag{6.15}$$

where $F(s)$ is an Abelian integral of the first kind. The analyticity of $F(s)$ on the surface $\hat{M}$ follows from (1.28). Since the left hand side of equation (6.15) is multi-valued with respect to s , the function $F(s)$ is also multi-valued and its periods along the cycles $K_j (j = 1,\dots,2g)$ are equal to (see (6.4))

$$\begin{aligned} F_{2j} &= u_{2j} - 2\pi i(u,w_j') - 2\pi(u,iw_j') , \\ F_{2j-1} &= u_{2j-1} , \quad j = 1,\dots,g , \end{aligned} \tag{6.16}$$

where $u_j (j = 1,\dots,2g)$ are the periods of the solution along the cycles $K_j (j = 1,\dots,2g)$.

If the solution has poles of the first kind, $u \in L_q^0(M)$, $1 < q < 2$, and hence the right hand side of (6.15) belongs to $L_q^0(M)$, too. It means that $F(q)$ is an Abelian integral of the second kind having poles of first order at the same points.

Analogous representations can be obtained also for solutions with poles of higher orders. For this purpose one can use measures having zeros at these points and corresponding weight spaces.

Regular solutions of the equation

$$\underset{\sim}{\partial}^* v \equiv -\bar{\partial} v + av + \overline{bv} = 0 \tag{6.17}$$

are representable in the form

$$v(s) - \frac{1}{\pi} \iint_{\hat{M}} [a(r)v(r) + \overline{b(r)}\,\overline{v(r)}] m(s,r)\, d\sigma_r = Z'(s) \tag{6.18}$$

where $Z'(s)$ is an Abelian covariant of the first kind. The pole at $s = q_0$ is a simple pole of the right hand side and by the residue theorem $Z'(s)$ is regular at this point.

Conversely, solutions of equations (6.15) and (6.18) are generalized analytic functions or covariants, regular or having poles.

C. The homogeneous equation

In the theory of generalized analytic functions in the complex plane $\mathbb{C}$ the following fact plays a principal role: the integral equation

$$u(z) - \frac{1}{\pi} \iint_{\mathbb{C}} \frac{a(t)u(t) + b(t)u(t)}{t-z}\, d\sigma_t = 0$$

has no nontrivial solutions (see § 1).

In this section we construct an example of an analogous equation having nontrivial solutions (Yu.L. Rodin [b]).

In § 9 the existence of such a covariant $a_0(s)$ of the type $(0,1)$ will be shown which is equal to zero outside of some simply-connected domain U such that the equation

$$\bar{\partial} u + a_0(s) u = 0 \tag{6.19}$$

has a solution $w_0(s)$ regular in the domain U and analytically continuable into the domain $M \setminus U$ and a multiple of the divisor $q_0 - \delta$, where δ, $\deg \delta = g$, is a characteristic divisor for the kernel $m_\delta(r,s)$.

Consider the equation

$$(1+P_\delta)u \equiv u(s) - \frac{1}{\pi}\iint_U a_0(r)\,\frac{w_0(s)}{\overline{w_0(s)}}\,\overline{u(r)}\,m_\delta(r,s)\,d\sigma_r = 0 \qquad (6.20)$$

If $w(s)$ is a solution of the equation

$$\bar{\partial}w + a_0(s)\,\frac{w_0(s)}{\overline{w_0(s)}}\,\overline{w(s)} = 0 \qquad (6.21)$$

regular in the domain U, analytic in $M \setminus U$ and a multiple of the divisor $q_0 - \delta$, then $w(s)$ is a solution of equation (6.20). Indeed, the function

$$F(s) = w(s) + P_\delta w(s)$$

is an analytic function on M which is a multiple of the divisor $-\delta$. As long as $\deg \delta = 1$, it follows that $F(s) \equiv$ const. Because $F(s)$ is equal to zero at q_0, we conclude that $F(s) \equiv 0$.

Now note that the function $w_0(s)$ possesses all these properties and hence is a nontrivial solution of equation (6.20).

§ 7. Generalized constants

A. The space L_0

Theorem 7.1 (Liouville). A regular solution of equation (6.14) on a closed Riemann surface M which is equal to zero at an arbitrary point $s_0 \in M$ is identically equal to zero.

Let $U \subset M$ be some domain on the surface, $s_0 \in U$ and $U' = M \setminus \bar{U}$ be the complement of its closure. Choose two single-valued kernels $m_U(r,s)$ and $m'_U(r,s)$ the characteristic divisors δ and δ' of which belong to the domains U and U', respectively*). Then the regular solution $u(s)$ is representable in the form

$$u(s) = \begin{cases} \varphi(s) \exp \frac{1}{\pi} \iint\limits_U [a(r)u(r) + b(r)\overline{u(r)}]\, m_U(r,s)\, d\sigma_r \\ \varphi'(s) \exp \frac{1}{\pi} \iint\limits_{U'} [a(r)u(r) + b(r)\overline{u(r)}]\, m'_U(r,s)\, d\sigma_r \end{cases} \tag{7.1}$$

where $\varphi(s)$ and $\varphi'(s)$ are holomorphic function in the domains U and U' respectively and $\varphi(s_0) = 0$. Assume φ and φ' are not identically zero. For $p \in \partial U$ we obtain the relation

$$\varphi(p) = \varphi'(p) \exp \left[\frac{1}{\pi} \iint\limits_{U'} [a(r)u(r) + b(r)\overline{u(r)}]\, m'_U(r,p)\, d\sigma_r - \right.$$

$$\left. - \frac{1}{\pi} \iint\limits_U [a(r)u(r) + b(r)\overline{u(r)}]\, m_U(r,p)\, d\sigma_r \right]. \tag{7.2}$$

Because the exponent on the right hand side of (7.2) is single-valued, the increments of the arguments of the functions $\varphi(p)$ and $\varphi'(p)$ have opposite signs and are equal to the common multiplicities of zeros of the function $\varphi(p)$ and $\varphi'(p)$ in the domains U and U', respectively,

---

*) A divisor $\delta = \sum_1^g q_k$ has to satisfy the single contition $\dim L(\delta) = 1$ or the equivalent condition $\dim H(\delta) = 0$. Choose $g - 1$ points in the domain U, $q_1,\dots,q_{g-1} \in U$. It is clear that there exists a single Abelian differential of the first kind dw such that $dw(q_j) = 0$, $j = 1,\dots,g-1$. As q_g an arbitrary point q can be chosen at which $dw(q) \neq 0$.

$$\frac{1}{2\pi}\Delta_{\partial U}\arg\varphi(p) = N_U(\varphi) = -\frac{1}{2\pi}\Delta_{\partial U}\arg\varphi'(p) =$$

$$= -N_{U'}(\varphi') : N_U + N_{U'} = 0 .$$

The last equality can be valid only for $N_U = N_{U'} = 0$. This contradicts the condition $u(s_0) = 0$.

Denote the space of regular solutions of equation (6.14) by L_0. Such solutions are called generalized constants. Because of the Liouville theorem, the real dimension of the space L_0, $l_o = \dim L_0$, is no more then two, $l_0 \leq 2$.

The case $l_0 = 2$ is realized for $a = b = 0$ (two constants, 1 and i). An example for $l_0 = 0$ is represented by the equation

$$\bar{\partial}u + a_0(s)\,u = 0 \tag{7.3}$$

for

$$\iint_{\hat{M}} a_0(s)dw_j(s) \neq 0 , \qquad j = 1,\ldots,g . \tag{7.4}$$

Briefly it is pointed out how to construct an example of $l_0 = 1$. Consider the equation

$$\bar{\partial}u + a_0u - a_0\bar{u} = 0 . \tag{7.5}$$

Since $u = 1$ is a solution of the equation (7.5), in this case $l_0 \leq 1$. As it will be shown in § 7B, $l_0 = 1$ if there exists a covariant of the type (1,0) satisfying the equation

$$\bar{\partial}v - a_0v + \overline{a_0v} = 0 \tag{7.6}$$

and having a single pole of first order on M. Note, that it is not closed.

Assuming

$$v = \xi + i\eta , \quad \bar{\partial} = \frac{1}{2}\left(\frac{\partial}{\partial x} + i\frac{\partial}{\partial y}\right) , \qquad z = x + iy ,$$

we obtain the real system

$$\frac{\partial\xi}{\partial x} - \frac{\partial\eta}{\partial y} = 0 ,$$
$$\frac{\partial\xi}{\partial y} + \frac{\partial\eta}{\partial x} = \alpha\xi + \beta\eta . \tag{7.7}$$

It means that the differential form $\eta dx + \xi dy$ is closed, i.e.

$$\eta dx + \xi dy = dV(s)$$

where $V(s)$ is a real (may be multi-valued) function on M. Choose as $V(s)$ the real function having a single logarithmic singularity of the form $\ln (x^2 + y^2)$, assume $v(s) = i\partial V(s)$ and define the function $a_0(s)$ by the condition

$$2i \operatorname{Im} [a_0(s)v(s)] = \bar{\partial} v .$$

This completes the construction of our example*) .

B. Calculation of the index of the operator $\underset{\sim}{\partial}$

Theorem 7.2. The space L_0 of generalized constants of the operator $\underset{\sim}{\partial}$ coincides with the subspace of the single-valued solutions of

$$Su = (1 + P)u = u(s) - \frac{1}{\pi} \iint_{\hat{M}} [a(r)u(r) + b(r)\overline{u(r)}]\, m(r,s)\, d\sigma_r = c \tag{7.8}$$

where c is a constant.

It is clear that any single-valued solution of equation (7.8) is a generalized constant. Conversely, let $u(s)$ be a generalized constant. Then the function $F(s) = Su$ is an Abelian integral of the first kind on M the periods of which along the cycles K_{2j-1} $(j = 1,\dots,g)$ are equal to zero (see (5.19)). Such an integral is a constant (O. Forster [a], G. Springer [a], R. Gunning [a]).

Consider the matrix

$$\Lambda = \begin{bmatrix} (u_1,w_1') & \dots (u_1,w_g') & (u_1,iw_1') & \dots (u_1,iw_g') \\ \\ (u_{g_0},w_1) & \dots (u_{g_0},w_g) & (u_{g_0},iw_1') & \dots (u_{g_0},iw_g') \end{bmatrix} \tag{7.9}$$

*) This example has been constructed by S.Ja. Gusman and the author.

where u_j $(j = 1,\dots,g_0)$ is a complete system of solutions of the equation $Su = 0$ and dw_l $(l = 1,\dots,g)$ is the basis (5.1) of the space of Abelian differentials of the first kind.

Note that

$$\operatorname{rank} \Lambda = g_0 . \tag{7.10}$$

In fact, if u is given by a zero linear combination of the lines of the matrix (7.9), then

$$(u,w_j') = (u,iw_j') = 0 , \qquad j = 1,\dots,g_0 . \tag{7.11}$$

Because of (6.16) this means that the function $u(s)$ is single-valued on M. By the normalization of the kernel, $u(q_0) = 0$. By virtue of the Liouville theorem, $u(s) \equiv 0$.

From equation (7.10) it follows, in particular, that $g_0 \leq 2g$.

The space $H(-q_0)$ is the space of generalized analytic covariants of the type (1,0) which are multiples of the divisor $-q_0$ (i.e. can possess a pole of first order at the point q_0).

Theorem 7.3. The space $H(-q_0)$ coincides with the space of solutions of the equation $S^*v = w'(s)$ where $w'(s)$ is an arbitrary Abelian covariant of the first kind. The dimension of this space is

$$\dim H(-q_0) = 2g . \tag{7.11'}$$

Indeed, if $v \in H(-q_0)$, then $v \in L_q^{1,0}(M)$, $1 < q < 2$, and hence $P^*v \in L_q^{1,0}(M)$, i.e. S^*v is an Abelian covariant having a pole at the point q_0 the order of which is not greater than 1. By the residue theorem, such a covariant is an Abelian covariant of the first kind. Conversely, if w' is an Abelian covariant of the first kind and the equation $S^*v = w'$ is solvable, then $v = w' - P^*v$ and hence the covariant v can have a pole only at the point q_0, i.e. $v \in H(-q_0)$.

Denote by $\hat{H}$ the space of Abelian covariants of the first kind satisfying the conditions

$$(u_j,w') = 0 , \qquad j = 1,\dots,g_0 \tag{7.12}$$

where u_j $(j = 1,\dots,g_0)$ is a basis of the space of solutions of the equation $Su = 0$.

By (7.10), $\dim \hat{H} = 2g - g_0$. A basis of the space $H(-q_0)$ consists of g_0 solutions of the equation $S^*v = 0$ and $2g - g_0$ solutions of the equation $S^*v = w' \in \hat{H}$. It involves equation (7.11').

The space H_0 of regular solutions of the equation $\underset{\sim}{\partial}^*v = 0$ called generalized analytic covariants of the first kind consists of those elements of the space $H(-q_0)$ which are regular at the point q_0 . From (6.12) it follows that for regularity of v at the point q_0 it is necessary and sufficient that

$$(1,v) = (i,v) = 0 \ . \tag{7.13}$$

Let $Z'_1,\ldots,Z'_{2g-g_0}$ be some basis of the space H of Abelian covariants of the first kind satisfying the conditions (7.12) and $v_1,\ldots,v_{2g}$ be a basis of the space $H(-q_0)$ such that $v_1,\ldots,v_{g_0}$ is a basis of the space of solutions of the equation $S^*v = 0$,

$$S^*v_j = 0 \ , \qquad j = 1,\ldots,g_0 \ , \tag{7.14}$$

and

$$S^*v_{j+g_0} = Z'_j \ , \qquad j = 1,\ldots,2g-g_0 \ . \tag{7.14'}$$

From (7.13) it follows that

$$\dim H_0 = 2g - \operatorname{rank} \Sigma_0 \tag{7.15}$$

where

$$\Sigma_0 = \begin{bmatrix} (1,v_1) & \ldots & (1,v_{2g}) \\ & & \\ (i,v_1) & \ldots & (i,v_{2g}) \end{bmatrix} . \tag{7.16}$$

Let us revert to the space L_0 of generalized constants of the operator $\underset{\sim}{\partial}$.

<u>Theorem 7.4</u>.

$$\dim L_0 = 2 - \operatorname{rank} \Sigma_0 \ . \tag{7.17}$$

Indeed, every generalized constant is a solution of the equation $Su = c$. This means that the solvability conditions for this equation

$$(c,v_1) = 0 \ , \ \ldots,(c,v_{g_0}) = 0 \tag{7.18}$$

are valid. This solution has to be single-valued. By (6.16) we have

$$(u,w_j') = 0 \ , \ (u,iw_j') = 0 \ , \qquad j = 1,\ldots,g \ . \tag{7.19}$$

From (7.19) it follows that for the Abelian covariants Z_k' $(k = 1,\ldots,2g-g_0)$ expressed by the basis $w_1',\ldots,w_g'$, $iw_1',\ldots,iw_g'$ the relations

$$(u,Z_j') = 0 \ , \qquad j = 1,\ldots,2g-g_0 \tag{7.20}$$

are valid. We obtain

$$0 = (u,Z_j') = (u,S^*v_{g_0+j}) = (Su,v_{g_0+j}) = (c,v_{g_0+j}) \ , \qquad j = 1,\ldots,2g-g_0 \ . \tag{7.21}$$

The relations (7.18) and (7.21) mean that

$$\dim L_0 \leq 2 - \operatorname{rank} \Sigma_0 \ . \tag{7.22}$$

Consider a vanishing linear combination of the lines of the matrix (7.16)

$$(c,v_j) = 0 \ , \qquad j = 1,\ldots,2g \ . \tag{7.23}$$

In virtue of the first g_0 relations (7.23), the equation $Su_0 = c$ is solvable. We have

$$0 = (c,v_{g_0+j}) = (Su_0,v_{g_0+j}) = (u_0,S^*v_{g_0+j}) = (u_0,Z_j') \ , \qquad j = 1,\ldots,2g-g_0 \ . \tag{7.24}$$

The set of solutions of the equation $Su = c$ consists of single-valued functions. In order to show this, construct the basis $Z_1',\ldots,Z_{2g}'$ of the space of Abelian differentials of the first kind. Here $Z_1',\ldots,Z_{2g-g_0}'$ are the covariants (7.14') satisfying the conditions (7.12) and the covariants $Z'_{2g-g_0+1},\ldots,Z'_{2g}$ are determined by the conditions

$$(u_i, Z'_j) = \delta_{ij}\ , \ i = 1,\ldots,g_0\ ;\ j = 2g-g_0+1,\ldots,2g \tag{7.25}$$

where u_i $(i = 1,\ldots,g_0)$ is a basis of the space of solutions of the equation $Su = 0$. The general solution of the equation $Su = c$ has the form

$$u = u_0 + \sum_{j=1}^{g_0} c_j u_j \tag{7.26}$$

where c_j are arbitrary real constants. We obtain from (6.16)

$$(u, w'_j) = (u, iw'_j) = 0\ , \qquad j = 1,\ldots,g\ . \tag{7.27}$$

Expressing the basis $(w'_j, iw'_j, j = 1,\ldots,g)$ via the basis Z'_k $(k = 1,\ldots,2g)$ we obtain

$$w'_j = \sum_{k=1}^{2g} \xi_{jk} Z'_k\ ,\ iw_j = \sum_{k=1}^{2g} \xi_{g+j,k} Z'_k\ , \qquad j = 1,\ldots,g \tag{7.28}$$

whence

$$\sum_{k=1}^{2g} \xi_{jk}\ (u, Z'_k) = 0\ , \qquad j = 1,\ldots,2g\ . \tag{7.29}$$

The relations

$$(u, Z'_k) = (u_0, Z'_k) + \sum_j c_j\ (u_j, Z'_k) = 0\ , \quad k = i,\ldots,2g - g_0\ . \tag{7.30}$$

follow from equation (7.12) and equation (7.24). We have

$$(u, Z'_k) = (u_0, Z'_k) + c_k = 0\ ,\ k = 2g-g_0+1,\ldots,2g$$

and hence

$$c_k = -\ (u_0, Z'_k)\ ,\ k = 2g-g_0+1,\ldots,2g\ .$$

Therefore

$$\dim L_0 \geq 2 - \operatorname{rank} \Sigma_0\ . \tag{7.31}$$

The inequalities (7.22) and (7.31) involve (7.17).

We obtain from (7.17) and (7.15)

Theorem 7.6 (Riemann-Roch).

$$\dim L_0 - \dim H_0 = 2 - 2g \,. \tag{7.32}$$

This equality expresses the index of the operator $\underset{\sim}{\partial}$ and concludes the complete proof of the Riemann-Roch theorem adduced in Chapter 1 [see Yu.L. Rodin [e,f,h], W. Koppelman [a]).

The relation (7.32) can be obtained by another method using the homotopy between the operators $\underset{\sim}{\partial}$ and $\bar{\partial}$. But our method can be used in more difficult situations when the index is infinite (K.L. Volkoviski [a]) or when the operators $\underset{\sim}{\partial}$ and $\bar{\partial}$ are not homotopic (A. Turakulov [a], [b]), see § 11.

Note that, as it follows from (7.15) and (7.17), the value rank Σ_0 is equal to the number of independent elements of the space $H(-q_0)$ with a pole at q_0 and causes a decrement of the number of generalized constants. This was used in section A to construct the example for $l_0 = 1$.

C. Pseudoanalytic functions

Consider until further notice the case $l_0 = 2$. In this case the equations $Su = 1$ and $Su = i$ possess single-valued solutions, w_1 and w_2, respectively, and any combination

$$u(s) = c_1 w_1(s) + c_2 w_2(s) \tag{7.33}$$

is a generalized constant, too. By the argument principle the function $u(s)$ may have zeros only if $c_1 = c_2 = 0$. Whence the inequality

$$\operatorname{Im}\,[\overline{w_1(s)}w_2(s)] \geq c_0 > 0 \,, \quad s \in M \,,\ c_0 \text{ a constant} \tag{7.34}$$

is valid. From

$$\bar{\partial}w_1 + aw_1 + b\bar{w}_1 = 0 \,,$$

$$\bar{\partial}w_2 + aw_2 + b\bar{w}_2 = 0$$

we obtain

$$a = \frac{\bar{w}_2\bar{\partial}w_1 - \bar{w}_1\bar{\partial}w_2}{w_2\bar{w}_1 - w_1\bar{w}_2}\ , \quad b = \frac{w_1\bar{\partial}w_2 - w_2\bar{\partial}w_1}{w_2\bar{w}_1 - w_1\bar{w}_2}\ . \tag{7.35}$$

Therefore, in the case $l_0 = 2$ there is a one-to-one correspondence between the generalized constants and the coefficients a and b.

The pair of functions (w_1, w_2), $w_1, w_2 \in C_{\frac{p-2}{p}}(M)$, $\bar{\partial}w_1, \bar{\partial}w_2 \in L_p^{0,1}(M)$ satisfying (7.34) is called the generating pair in the sense of L. Bers (L. Bers [a,b,c], L. Bers, Nirenberg [a,b]). Every single-valued function on **M** is representable in the form

$$u(s) = \chi_1(s)w_1(s) + \chi_2(s)w_2(s) \tag{7.36}$$

where χ_1, χ_2 are real.

The derivation in the sense of L. Bers in defined by the relation

$$\dot{u}(z(s_0)) = \lim_{s\to s_0} \frac{u(z(s)) - \chi_1(z(s_0))w_1(z(s)) - \chi_2(z(s_0))w_2(z(s))}{z(s) - z(s_0)}\ . \tag{7.37}$$

As in the case of the complex plane, at the point $z(s_0)$ the relation

$$\bar{\partial}u(s_0) + a(s_0)u(s_0) + b(s_0)\overline{u(s_0)} = 0 \tag{7.38}$$

where the coefficients $a(s_0)$ and $b(s_0)$ are determined by the formulae (7.35), is valid.

L. Bers calls the function $u(s)$ a pseudoanalytic function of the first kind. The function $\chi = \chi_1 + i\chi_2$ is called a pseudoanalytic function of the second kind. It satisfies the equation

$$\bar{\partial}\chi - q(s)\overline{\partial\chi} = 0\ , \quad q = \frac{w_1 + iw_2}{w_1 - iw_2}\ . \tag{7.39}$$

From (7.34) it follows that the system (7.39) is elliptic and

$$|q(s)| \le q_0 < 1 \quad \text{on} \quad M\ . \tag{7.40}$$

It means that the function $\chi = \chi(s)$ realizes a quasi-conformal mapping of the surface M.

§ 8. The Riemann-Roch theorem

In this section we adduce the direct proof of the theorem of Riemann-Roch using the same idea as in § 7 (Yu.L. Rodin [e,f]).

Let $p_1,\dots,p_n$ be different points of M and $\gamma = \sum_{j=1}^{n} p_j$. A multi-valued solution which is a multiple of the divisor $-\gamma$ with zero periods along the cycles K_{2j-1} $(j = 1,\dots,g)$ is a solution of the equation

$$Su = t_\gamma(s) \ , \ t_\gamma(s) = \alpha_0 + i\alpha_1 + \sum_{k=1}^{n} (-\alpha_{2k} + i\alpha_{2k+1})\, t_{p_k}(s) \tag{8.1}$$

where $t_{p_k}(s)$ are the Abelian integrals of the second kind (5.5) and α_k $(k = 0,\dots,2n+1)$ are real numbers.

Let $v_1,\dots,v_{2g}$ be the basis of the space $H(-q_0)$ introduced in § 7 (see (7.14), (7.14'))

$$\begin{aligned} S^*v_j &= 0 \quad , \quad j = 1,\dots,g_0 \ , \\ S^*v_{j+g_0} &= Z_j' \ , \quad j = 1,\dots,2g-g_0 \ . \end{aligned} \tag{8.2}$$

Consider the matrix Σ_γ (figure 7).

<u>Theorem 8.1</u>. The real dimension of the space $H(\gamma)$ of generalized analytic covariants of the first kind, i.e. regular solutions of the equation $\underset{\sim}{\partial}^*v = 0$ which are multiples of the divisor γ is equal to

$$\dim H(\gamma) = 2g - \operatorname{rank} \Sigma_\gamma \ . \tag{8.3}$$

Let

$$v \in H(\gamma) \subset H(-q_0) \ , \ v = \sum_{j=1}^{2g} c_j v_j \ .$$

Then the covariant $v(s)$ is regular at the points p_k $(k = 1,\dots,n)$ and satisfies the equation

$$S^*v = \sum_{j=1}^{2g-g_0} c_{j+g_0} Z_j' = Z' \ . \tag{8.4}$$

We obtain the relations

$$(1,v) = (i,v) = 0 \;,$$

$$-(t_{p_j},v) + \sum_{k=1}^{2g-g_0} c_{g_0+k} \operatorname{Re} Z_k'(p_j) = 0 \;,$$

$$(it_{p_j},v) + \sum_{k=1}^{2g-g_0} c_{g_0+k} \operatorname{Im} Z_k'(p_j) = 0 \;. \tag{8.5}$$

These relations follow from the relations

$$v(p_j) = \frac{1}{\pi} \iint_{\hat{M}} [a(r)v(r) + \overline{b(r)}\;\overline{v(r)}]\, m(p_j,r)\, d\sigma_r + Z'(p_j) \tag{8.6}$$

and (see (5.26))

$$m(p_j,r) = - t_{p_j}(r) \;. \tag{8.7}$$

Therefore, every element of the space $H(\gamma)$ corresponds to a vanishing linear combination of the columns of the matrix Σ_γ . Conversely, let us have a vanishing linear combination of the columns of the matrix Σ_γ with coefficients c_j $(j = 1,\dots,2g)$. Then the covariant $v = \sum_{j=1}^{2g} c_j v_j$ satisfies the conditions (8.5) and hence $v \in H(\gamma)$.

Theorem 8.2. The real dimension of the space $L(\gamma)$ of generalized analytic functions which are multiples of the divisor $-\gamma$ is equal to

$$\dim L(\gamma) = 2 \deg \gamma + 2 - \operatorname{rank} \Sigma_\gamma \;. \tag{8.8}$$

$$
\begin{bmatrix}
(1,v_1) & \dots & (1,v_{g_0}) & (1,v_{g_0+1}) & \dots & (1,v_{2g}) \\
(i,v_1) & \dots & (i,v_{g_0}) & (i,v_{g_0+1}) & \dots & (i,v_{2g}) \\
-(t_{p_1},v_1) & \dots & -(t_{p_1},v_{g_0}) & -(t_{p_1},v_{g_0+1}) + \mathrm{Re}\, Z_1'(p_1) & \dots & -(t_{p_1},v_{2g}) + \mathrm{Re}\, Z_{2g-g_0}'(p_1) \\
(it_{p_1},v_1) & \dots & (it_{p_1},v_{g_0}) & (it_{p_1},v_{g_0+1}) + \mathrm{Re}\, Z_1'(p_1) & \dots & (it_{p_1},v_{2g}) + \mathrm{Im}\, Z_{2g-g_0}'(p_1) \\
\cdots & & \cdots & \cdots & \cdots & \cdots \\
-(t_{p_n},v_1) & \dots & -(t_{p_n},v_{g_0}) & -(t_{p_n},v_{g_0+1}) + \mathrm{Re}\, Z_1'(p_n) & \dots & -(t_{p_n},v_{2g}) + \mathrm{Re}\, Z_{2g-g_0}'(p_n) \\
(it_{p_n},v_1) & \dots & (it_{p_n},v_{g_0}) & (it_{p_n},v_{g_0+1}) + \mathrm{Im}\, Z_1'(p_n) & \dots & (it_{p_n},v_{2g}) + \mathrm{Im}\, Z_{2g-g_0}'(p_n)
\end{bmatrix}
$$

The matrix Σ_γ

Figure 7

Let some zero combinations of the lines of the matrix Σ_γ correspond to the coefficients $\alpha_0,\dots,\alpha_{2n+1}$ and

$$t_\gamma(s) = (\alpha_0 + i\alpha_1) + \sum_{k=1}^{n} (-\alpha_{2k} + i\alpha_{2k+1})\, t_{p_k}(s) \,. \tag{8.9}$$

Then

$$(t_\gamma, v_j) = 0 \,, \qquad j = 1,\dots,g_0 \,, \tag{8.10}$$

$$(t_\gamma, v_{g_0+j}) - \mathrm{Re} \sum_{k=1}^{n} (-\alpha_{2k} + i\alpha_{2k+1})\, Z_j'(p_k) = 0 \,, \qquad j = 1,\dots,2g-g_0 \,. \tag{8.11}$$

From (8.10) it follows that the equation $Su = t_\gamma$ is solvable. Let u_0 be a solution. Then (8.11) can be transformed into

$$0 = (Su_0, v_{g_0+j}) - \mathrm{Re} \sum_{k=1}^{n} (-\alpha_{2k} + i\alpha_{2k+1})\, Z_j'(p_k)$$

$$= (u_0, S^* v_{g_0+j}) - \mathrm{Re} \sum_{k=1}^{n} (-\alpha_{2k} + i\alpha_{2k+1})\, Z_j'(p_k) =$$

$$= (u_0, Z_j') - \mathrm{Re} \sum_{k=1}^{n} (-\alpha_{2k} + i\alpha_{2k+1}) \; Z_j'(p_k) \; , \qquad (8.12)$$

$$j = 1, \ldots, 2g - g_0 \; .$$

The relations (8.12) are valid for any solution of the equation $Su = t_\gamma$ since for any solution of the equation $Su_k = 0$, $k = 1, \ldots, g_0$, we have

$$(u_k, Z_j') = (u_k, S^* v_{g_0+j}) = (Su_k, v_{g_0+j}) = 0 ,$$

$$k = 1, \ldots, g_0 \; , \; j = 1, \ldots, 2g - g_0 \; .$$

Add the set $\{Z_j' \; , \; j = 1, \ldots, 2g-g_0\}$ to the basis (7.25). Then there exists such a solution u of the equation $Su = t_\gamma$ that

$$u = u_0 + \sum_{k=1}^{g_0} c_k u_k$$

and

$$(u, Z_j') - \mathrm{Re} \sum_{k=1}^{n} (-\alpha_{2k} + i\alpha_{2k+1}) \; Z_j'(p_k) = 0 \; , \qquad (8.13)$$

$$j = 2g - g_0 + 1, \ldots, 2g \; .$$

By virtue of (7.25) we conclude that the relation

$$(u, w') - \mathrm{Re} \sum_{k=1}^{n} (-\alpha_{2k} + i\alpha_{2k-1}) w'(p_k) = 0$$

is valid for any Abelian covariant of the first kind. In particular, for the basis (5.1) dw_j , idw_j , $j = 1, \ldots, g$ we have

$$(u, w_j') - \mathrm{Re} \sum_{k=1}^{n} (-\alpha_{2k} + i\alpha_{2k+1}) \; w_j'(p_k) = 0 \; ,$$

$$(8.14)$$

$$(u, iw_j') - \mathrm{Re} \sum_{k=1}^{n} (-\alpha_{2k} + i\alpha_{2k+1}) \; iw_j'(p_k) = 0 \; , \quad j = 1, \ldots, g \; .$$

Because of (5.9),

$$- 2\pi i \sum_{k=1}^{n} (-\alpha_{2k} + i\alpha_{2k+1}) \; w_j'(p_k) = \int_{K_{2j}} dt_\gamma \; . \qquad (8.15)$$

Then, from (6.16) it follows that

$$\operatorname{Re} u_{2j} = 2\pi(u,iw_j') + \operatorname{Re} \int_{K_{2j}} dt_\gamma = 0 \,,$$
$$\operatorname{Im} u_{2j} = 2\pi(u,w_j') + \operatorname{Im} \int_{K_{2j}} dt_\gamma = 0 \,. \tag{8.16}$$

It means that the function $u(s)$ is single-valued. Therefore, any linear combination of lines of Σ_γ corresponds to an element of the space $L(\gamma)$, i.e.

$$\dim L(\gamma) \geq 2 \deg \gamma + 2 - \operatorname{rank} \Sigma_\gamma \,. \tag{8.17}$$

Let now $u_0 \in L(\gamma)$. Then the function $u_0(s)$ is single-valued on M and satisfies the integral equation

$$Su_0 = t_\gamma(s) \,, \quad t_\gamma(s) = (\alpha_0 + i\alpha_1) + \sum_{k=1}^{n} (-\alpha_{2k} + i\alpha_{2k+1})\, t_{p_k}(s) \,. \tag{8.18}$$

The solvability of the equation (8.18) means that the conditions

$$(t_\gamma, v_j) = 0 \,, \quad j = 1,\dots,g_0 \tag{8.19}$$

are valid. Here v_j $(j = 1,\dots,g_0)$ are solution of the equation $S^*v = 0$.

Further, the single-valuedness of the function $u_0(s)$ means, in virtue of (6.16), that

$$-2\pi i(u_0,w_j') - 2\pi(u_0,iw_j') = \int_{K_{2j}} dt_\gamma \,, \quad j = 1,\dots,g \,. \tag{8.20}$$

By (5.9) we obtain

$$-i(u_0,w_j') - (u_0,iw_j') = -\sum_{k=1}^{n} i(-\alpha_{2k} + i\alpha_{2k+1})\, w_j'(p_k) \,, \quad j = 1,\dots,g \,,$$

what is equivalent to the relations

$$(u_0, w_j') + \sum_{k=1}^{n} (\alpha_{2k} \operatorname{Re} w_j'(p_k) + \alpha_{2k+1} \operatorname{Im} w_j'(p_k)) = 0 ,$$

$$(u_0, iw_j') + \sum_{k=1}^{n} (\alpha_{2k} \operatorname{Re} iw_j'(p_k) + \alpha_{2k+1} \operatorname{Im} iw_j'(p_k)) = 0 , \quad j = 1,\dots,g . \tag{8.21}$$

Hence it appears that the relation

$$(u_0, w') + \sum_{k=1}^{n} (\alpha_{2k} \operatorname{Re} w'(p_k) + \alpha_{2k+1} \operatorname{Im} w'(p_k)) = 0$$

is valid for any Abelian differential of the first kind.

In particular, assume $w' = Z_j'(s)$, $j = 1,\dots,2g-g_0$. We have

$$(u_0, Z_j') + \sum_{k=1}^{n} (\alpha_{2k} \operatorname{Re} Z_j'(p_k) + \alpha_{2k+1} \operatorname{Im} Z_j'(p_k)) =$$

$$(u_0, S^* v_{g_0+j}) + \sum_{k=1}^{n} (\alpha_{2k} \operatorname{Re} Z_j'(p_k) + \alpha_{2k+1} \operatorname{Im} Z_j'(p_k)) = \tag{8.22}$$

$$(t_\gamma, v_{g_0+j}) + \sum_{k=1}^{n} (\alpha_{2k} \operatorname{Re} Z_j'(p_k) + \alpha_{2k+1} \operatorname{Im} Z_j'(p_k)) = 0 .$$

It means that the combination of lines of the matrix Σ_γ with coefficients α_j $(j = 0,\dots,2n+1)$ is equal to zero.

From Theorem 8.1 and Theorem 8.2 it follows

Theorem 8.3 (Riemann-Roch).

$$\dim L(\gamma) - \dim H(\gamma) = 2 \deg \gamma - 2g + 2 . \tag{8.23}$$

The relation (8.23) has been proved for a positive divisor without multiple points.

Let now γ be an arbitrary divisor. Choose a positive divisor γ_0 without multiple points such that $\deg \gamma = \deg \gamma_0$. As it will be shown in § 9 the coefficient $a_0(s)$ may be choosen such that the equation

$$\bar{\partial} u + a_0(s) u = 0 \tag{8.24}$$

has a solution $u_0(s)$ determined by the divisor $(u_0) = \gamma - \gamma_0$.

The divisor $\gamma - \gamma_0 = \Sigma\, \tilde{\alpha}_k q_k$ has zero degree, $\deg(\gamma - \gamma_0) = \Sigma\, \tilde{\alpha}_k = 0$. The coefficient $a_0(s)$ has to satisfy the conditions

$$\iint_{\hat{M}} a_0(s) w_j'(s)\, d\sigma_s \equiv \sum_k \alpha_k w_j(q_k) \quad (\text{mod periods of } w_j) \,. \tag{8.25}$$

Let $L(\gamma)$ be the space of generalized analytic functions which are multiples of the divisor $-\gamma$ for the operator $\underset{\sim}{\partial}$ and $H(\gamma)$ be the space of generalized analytic covariants which are multiples of the divisor γ for the operator $\underset{\sim}{\partial}^*$. Let $u \in L(\gamma)$ and $v \in H(\gamma)$. Then $uu_0 \in L_0(\gamma_0)$ and $\frac{v}{u_0} \in H_0(\gamma_0)$ where the space $L_0(\gamma_0)$ and $H_0(\gamma_0)$ correspond to the operators $\underset{\sim}{\partial}_0$ and $\underset{\sim}{\partial}_0^*$ with coefficients

$$a_0(s) = a(s) \,, \; b_0(s) = b(s) \frac{u_0(s)}{\overline{u_0(s)}} \,. \tag{8.26}$$

Therefore, we obtain the Riemann-Roch theorem for arbitrary divisors.

CHAPTER 3

THE RIEMANN BOUNDARY PROBLEM

§ 9. The Riemann boundary problem

A. The equation $\bar{\partial}u + a(s)u = 0$

1. First of all, consider the inhomogeneous Cauchy-Riemann equation

$$\bar{\partial}u + a(s) = 0 \qquad a \in L_p^{0,1}(M) \ , \quad p > 2 \ . \tag{9.1}$$

It is clear that an arbitrary, single-valued or multi-valued solution of equation (9.1) is representable in the form

$$u(s) = \frac{1}{\pi} \iint_{\hat{M}} a(p)m(p,s)\, d\sigma_p + \Phi(s) \tag{9.2}$$

where $\Phi(s)$ is an analytic function (more exactly, an Abelian integral).

For equation (9.1) it is naturally to consider the Plemelj-Sokhotskii boundary problem. Let L be a contour on M with an H-continuously varying tangent consisting of $m+1$ components, $L = \sum_{j=0}^{m} L_j$, $m \geq 0$, $L_i \cap L_j = \emptyset$ if $i \neq j$. Every curve L_j is assumed to be closed and without self-intersections. The contour L has to divide the surface into two domains $T^{\pm}$. Let the function $g(p)$ be determined on the contour L , $p \in L$, satisfying a Hölder condition. It is necessary to determine solutions $u^{\pm}(s)$ of the equation (9.1) regular in $T^{\pm}$, continuous up to the boundary and satisfying on L the Plemelj-Sokhotskii boundary condition

$$u^{+}(p) - u^{-}(p) = g(p) \ , \qquad p \in L \ . \tag{9.3}$$

Note. If the contour L does not divide the surface M (for example, if L is a cyclic section of a torus), one may add a dividing contour L' (i.e. construct a parallel cyclic section) and assume $g \equiv 0$ on L' .

Consider the Cauchy type integral

$$F(s) = \frac{1}{2\pi i} \int_L g(p)m(p,s)\,dz(p) \ . \tag{9.4}$$

Since the kernel of this integral has a pole

$$\frac{1}{z(p)-z(s)} \, ,$$

the integral (9.4) possesses all usual properties of Cauchy type integrals: the Cauchy integral formula is valid, the function (9.4) is holomorphic in $M \setminus L$ and is continuous up to the boundary, the Plemelj-Sokhotskii formulae

$$F^{\pm}(s) = \pm \frac{1}{2} g(s) + \frac{1}{2\pi i} \int_L g(p)m(p,s)dz(p) \, , \qquad s \in L \tag{9.5}$$

are valid. Here the integral is understood in the principal value sense. However, by the multi-valuedness of the kernel, the integral (9.4) is multi-valued on M and its periods along the cycles K_{2j} $(j = 1,\ldots,g)$ are equal to (see (5.19))

$$F_j = \int_L g(p)dw_j(p) \qquad j = 1,\ldots,g \, . \tag{9.6}$$

Substituting (9.5) into (9.2) we obtain the solution of the problem (9.3) in the form

$$u^0(s) = \frac{1}{\pi} \iint_{\hat{M}} a(p)m(p,s)\, d\sigma_p + \frac{1}{2\pi i} \int_L g(p)m(p,s)dz(p) \, . \tag{9.7}$$

The periods of this solution are equal to

$$u_j^0 = 2i \iint_{\hat{M}} a(p)w_j'(p)\, d\sigma_p + \int_L g(p)w_j'(p)dz(p) \, . \tag{9.8}$$

The general solution has the form

$$u(s) = u^0(s) + w(s)$$

where $w(s)$ is an Abelian integral of the first kind. If the solution is single-valued, $w(s) \equiv \text{const}$ since all periods of this integral along the cycles K_{2j-1} $(j = 1,\ldots,g)$ are equal to zero. Therefore, we obtain the solvability conditions for the Plemelj-

Sokhotskii problem in the class of singel-valued solutions of the equations (9.1) as

$$2i \iint_{\hat{M}} a(p)w_j'(p)\, d\sigma_p + \int_L g(p)w_j'(p)dz(p) = 0 , \quad j = 1,\dots,g . \tag{9.9}$$

2. We now pass over to the equation (see Yu.L. Rodin [d])

$$\bar{\partial}u + a(s)u = 0 , \; a \in L_q^{0,1}(M) , \; q > 2 . \tag{9.10}$$

We obtain the integral representation for the solutions

$$u(s) = \varphi(s) \exp \frac{1}{\pi} \iint_{\hat{M}} a(p)m(p,s)\, d\sigma_p \tag{9.11}$$

where $\varphi(s)$ is analytic (the exponent of an Abelian integral).

Consider the existence conditions for a solution having no poles (and, consequently, by the argument principle having no zeros). In this case the value $\ln \varphi(s)$ is an Abelian integral of the first kind the periods of which along the cycles K_{2j-1} $(j = 1,\dots,g)$ are equal to $-2\pi i n_j$ (n_j are integers) and, therefore,

$$\varphi(s) = \exp \{-2\pi i \sum_{j=1}^{g} n_j w_j(s)\} . \tag{9.12}$$

We obtain the single-valuedness conditions for the function (9.11) in equating its periods along the cycles K_{2l} $(l = 1,\dots,g)$ to $2\pi i n_l'$ with arbitrary integers n_l' $(l = 1,\dots,g)$,

$$2i \iint_{\hat{M}} a(p)w_l'(p)\, d\sigma_p - 2\pi i \sum_{j=1}^{g} n_j \int_{K_{2l}} dw_j(s) = 2\pi i n_l' .$$

Taking into account (5.1) and (5.1') we obtain the conditions

$$\frac{1}{\pi} \iint_{\hat{M}} a(p)w_l'(p)\, d\sigma_p = \sum_{j=1}^{g} (n_j \int_{K_{2j}} dw_l + n_j' \int_{K_{2j-1}} dw_l) = \int_K dw_l ,$$
$$K = \sum_{j=1}^{g} (n_j K_{2j} + n_j' K_{2j-1}) , \; l = 1,\dots,g . \tag{9.13}$$

The expression (9.13) is symbolically written in the form of the comparison system

$$\frac{1}{\pi}\iint_{\hat{M}} a(p)w_l'(p)\,d\sigma_p \equiv 0 \text{ (mod periods of } w_l) \quad l = 1,\dots,g . \tag{9.14}$$

Consider the Riemann boundary problem: determine the regular solutions $u^{\pm}(s)$ of equation (9.10) in the domains $T^{\pm}$ continuous up to the boundary and satisfying the boundary condition

$$u^{+}(p) = G(p)u^{-}(p) , \qquad p \in L \tag{9.15}$$

where $G(p)$ is an H-continuous function on L different from zero everywhere (see F.D. Gakhov [a], N.I. Muskhelishvili [a]) .

Consider the case where the function $\ln G(p)$ is single-valued on all contours L_j $(j = 0,\dots,m)$

$$\frac{1}{2\pi}\Delta_{L_j} \arg G(p) = 0 , \qquad j = 0,\dots,m . \tag{9.16}$$

Substituting the expression

$$\varphi(s) = \exp\left\{\frac{1}{2\pi i}\int_L \ln G(p)m(p,s)dz(p) - 2\pi i\sum_{j=1}^{g} n_j w_j(s)\right\} \tag{9.17}$$

into (9.11) we obtain the conditions for single-valuedness in the form

$$2i\iint_{\hat{M}} a(p)w_l'(p)d\sigma_p + \int_L \ln G(p)w_l'(p)dz(p) -$$

$$- 2\pi i\sum_{j=1}^{g} n_j \int_{K_{2l}} dw_j = 2\pi i\, n_l' , \qquad l = 1,\dots,g.$$

Transforming this expression as above, we yield the conditions

$$\frac{1}{\pi}\iint_{\hat{M}} a(p)w_l'(p)\,d\sigma_p + \frac{1}{2\pi i}\int_L \ln G(p)dw_l(p) \equiv 0 \quad \text{(mod periods of } w_l) ,$$

$$l = 1,\dots,g . \tag{9.18}$$

B. The Riemann problem for the equation $\bar{\partial}u + au = 0$*)

Consider the Riemann problem (9.15) for nonzero indices

$$\kappa_j = \frac{1}{2\pi} \Delta_{L_j} \arg G \;, \; \kappa = \sum_{j=0}^{m} \kappa_j \;. \tag{9.19}$$

On every curve L_j we fix a point p_j $(j = 0,\ldots,m)$. In this case the function

$$\Phi(s) = \exp \frac{1}{2\pi i} \int_L \ln G(p)m(p,s)dz(p) \tag{9.20}$$

is analytic, satisfies the boundary condition (9.15), has zeros of orders κ_j at the points p_j $(j = 0,\ldots,m)$ (if $\kappa_j < 0$, it is a pole of the order $|\kappa_j|$) , and multiplicative periods

$$\Phi_j = \exp \int_L \ln G(p)dw_j(p) \tag{9.21}$$

along the cycles K_{2j} $(j = 1,\ldots,g)$. We obtain the solution in the form

$$u(s) = \exp \{\frac{1}{\pi} \iint_{\hat{M}} a(p)m(p,s)\, d\sigma_p + \frac{1}{2\pi i} \int_L \ln G(p)m(p,s)dz(p) +$$

$$+ \sum_{l=0}^{m} \kappa_l \omega_{q_0 p_l}(s) + \sum_{j=1}^{\kappa} \omega_{q_j q_0}(s) - 2\pi i \sum_{l=1}^{g} n_l w_l(s)\} \tag{9.22}$$

where $\omega_{q_0 p}(s)$ are the Abelian integrals of the third kind (5.6), q_0 is an arbitrary fixed point of the surface and q_j $(j = 1,\ldots,\kappa)$ are points unknown for the present.

The periods of the function $\ln u(s)$ along cycles K_{2j-1} are equal to $-2\pi i\, n_j$ $(j = 1,\ldots,g)$. The periods along the cycles K_{2j} $(j = 1,\ldots,g)$ are equal to

---

*) See Yu.L. Rodin [a,c,d], W. Koppelman [b,c], L.I. Chibrikova [a], R.N. Abdulaev [a,b].

$$\ln u_j = \int_{K_{2j}} d \ln u = 2i \iint_{\hat{M}} a(p) w_j'(p)\, d\sigma_p + \int_L \ln G(p)\, dw_j +$$

$$+ \sum_{l=0}^{m} \kappa_l \int_{K_{2j}} d\omega_{q_0 p_l} + \sum_{k=1}^{\kappa} \int_{K_{2j}} d\omega_{q_k q_0} - 2\pi i \sum_{l=1}^{g} n_l \int_{K_{2j}} dw_l \,, \tag{9.23}$$

$$j = 1,\ldots,g \,.$$

Taking into account (5.1') and (5.8) we obtain the condition for single-valuedness in the form

$$\sum_{\alpha=1}^{\kappa} w_j(q_\alpha) = \sum_{l=0}^{m} \kappa_l w_j(p_l) + \frac{1}{2\pi i} \int_L \ln G\, dw_j +$$

$$+ \frac{1}{\pi} \iint_{\hat{M}} a(p) w_j'(p)\, d\sigma_p - \sum_{l=1}^{g} \left(n_l \int_{K_{2l}} dw_j + n_l' \int_{K_{2l-1}} dw_j\right) , \tag{9.24}$$

$$j = 1,\ldots,g \,.$$

Therefore, for the existence of the solution it is necessary and sufficient that there exist κ points $q_1,\ldots,q_\kappa$ satisfying the condition

$$\sum_{\alpha=1}^{\kappa} w_j(q_\alpha) = F_j \quad (\text{mod periods of } w_j) \,, \quad j = 1,\ldots,g \tag{9.25}$$

where

$$F_j = \sum_{l=1}^{m} \kappa_l w_j(p_l) + \frac{1}{2\pi i} \int_L \ln G\, dw_j + \frac{1}{\pi} \iint_{\hat{M}} a(p) w_j'(p)\, d\sigma_p \,, \tag{9.26}$$

$$j = 1,\ldots,g \,.$$

The problem (9.25) is known as the Jacobi inverse problem for Abelian integrals. In the following section we shall show that this problem is solvable for any F_j $(j = 1,\ldots,g)$ if $\kappa \geq g$.

The formula for the solution of the Riemann problem for analytic functions is adduced in the paper Yu.L. Rodin [o]. The solution is expressed by the Riemann theta function. In the considered case it is necessary to add the factor

$$\exp \frac{1}{\pi} \iint_{\hat{M}} a(p)m(p,s)\, d\sigma_p$$

and instead of

$$\frac{1}{2\pi i} \int_L \ln G \, dw_j$$

it is necessary to use the values

$$\frac{1}{2\pi i} \int_L \ln G \, dw_j + \frac{1}{\pi} \iint_{\hat{M}} a(p) w_j'(p)\, d\sigma_p \,, \quad j = 1,\dots,g \,.$$

Note that if instead of regular solutions we are looking for solutions which are multiples of the present divisor $-\gamma$,

$$(u) + \gamma \geq 0 \,,$$

this is equivalent to the problem of the index $\kappa' = \kappa + \deg \gamma$. Indeed, represent the solution in the form

$$u(s) = \begin{cases} \gamma^+(s) u_0^+(s) \,, & s \in T^+ \,, \\ \gamma^-(s) u_0^-(s) \,, & s \in T^- \,. \end{cases} \tag{9.27}$$

Here the functions $\gamma^{\pm}(s)$ are analytic functions in $T^{\pm}$ determined by the divisors $\gamma^{\pm} = \gamma \cap T^{\pm}$ (for simplicity, we assume that the points of γ do not belong to L). We obtain the boundary problem

$$u_0^+(p) = \frac{\gamma^-(p)}{\gamma^+(p)} G(p) u_0^-(p) \quad \text{on } L \tag{9.28}$$

of the index $\kappa + \deg \gamma$.

Since the Riemann problem is solvable for $\kappa \geq g$, it follows that any Riemann problem of index $\kappa < g$ has a solution possessing no more then $g - \kappa$ poles.

Let now $u_0(s)$ be any solution of the problem (9.15) in general having poles. Then a regular solution of the problem (9.15) is given in the form

$$u(s) = u_0(s) f(s) , \quad (f) \geq (u_0)$$

where $f(s)$ is an analytic function. Therefore, the number of solutions of the problem (9.15) is determined by the dimension $\dim L_0(u_0)$ of the space of analytic functions which are multiples of the divisor $-(u_0)$.

The conjugate problem

$$v^+(p) = \frac{1}{G(p)} v^-(p) \qquad p \in L \tag{9.29}$$

is raised for covariants of the type (1.0) satisfying the equation

$$\bar{\partial} v - a(s) v = 0 . \tag{9.30}$$

It is clear that every solution of this problem corresponds to an element of the space $H(u_0)$ of Abelian covariants which are multiples of the divisor (u_0). If l and h are numbers of solutions of the problem (9.15) and (9.29), respectively (the dimensions are real), then in virtue of the Riemann-Roch theorem we obtain the relation

$$l - h = 2\kappa - 2g + 2 \tag{9.31}$$

since by the argument principle the difference between the numbers of zeros and poles of the solution u_0, $\deg(u_0)$, is equal to κ.

Consider until further notice the case $\kappa = 0$. Instead of the inverse problem (9.25), we obtain the conditions

$$F_j \equiv 0 \ (\text{mod periods of } w_j) , \quad j = 1,\dots,g . \tag{9.31'}$$

Let, in particular, $G \equiv 1$ and let us look for the solution determined by the divisor

$$\gamma = \sum_l \alpha_l q_l , \ \deg \gamma = \sum_l \alpha_l = 0 .$$

We obtain the solution in the form

$$u(s) = \exp \{ \frac{1}{\pi} \iint_{\hat{M}} a(p) m(p,s) \, d\sigma_p +$$

$$+ \sum_l \alpha_l \omega_{p_0 q_l}(s) + 2\pi i \sum_{k=1}^{g} n_k w_k(s) \} \tag{9.32}$$

The conditions for single-valuedness for this expression have the form

$$\sum_{l} \alpha_l w_j(q_l) \equiv -\frac{1}{\pi} \iint_{\hat{M}} a(p) w_j'(p)\, d\sigma_p \quad (\text{mod periods of } w_j)\,,$$

$$j = 1,\ldots,g\,. \tag{9.33}$$

The system (9.33) gives the necessary and sufficient conditions for the existence of the solution to the problem (9.15) with prescribed zeros and poles (Abel's theorem).

The system (9.33) may be interpreted otherwise. Let $\gamma = \sum \alpha_s q_s$ be some divisor, $\deg \gamma = 0$. Select a coefficient $a(s) \in L_p^{0,1}(M)$, $p > 2$, satisfying the conditions (9.33). Therefore, for every divisor of zero degree there exists a coefficient $a(s)$ such that the equation (9.15) has a solution determined by this divisor. For the conditions (9.25) the same is valid.

C. The Jacobi inversion problem

Consider now the problem to determine the points $q_1,\ldots,q_\kappa$ satisfying the system

$$\sum_{k=1}^{\kappa} w_j(q_k) = F_j \quad (\text{mod periods of } w_j)\,,\ j = 1,\ldots,g \tag{9.34}$$

where

$$w_j(q) = \int_{q_0}^{q} dw_j\,, \quad j = 1,\ldots,g\,.$$

In virtue of the formulae (5.1), (5.1'), (5.8), and (9.13) the problem (9.34) can be rewritten into the form

$$\frac{1}{2\pi i} \sum_{k=1}^{\kappa} \int_{K_{2j}} d\omega_{q_0 q_k}(s) = F_j + \sum_{l=1}^{g} [n_l \int_{K_{2l}} dw_j + n_l' \int_{K_{2l-1}} dw_j] =$$

$$= F_j + \sum_{l=1}^{g} (n_l \int_{K_{2j}} dw_l + n_l')\,, \quad j = 1,\ldots,g\,,$$

where n_l, n_l' are integers. It means that if q_l $(l = 1,\ldots,\kappa)$ is a solution of the inversion problem then the value $\exp \omega(s)$,

$$\omega(s) = \sum_{k=1}^{\kappa} \omega_{q_0 q_k}(s) - 2\pi i \sum_{l=1}^{g} n_l w_l(s) , \tag{9.35}$$

is an analytic function having a zero of order κ at the point q_0, being a multiple of the divisor $-\gamma$, $\gamma = \sum_{k=1}^{\kappa} q_k$, and posessing the multiplicative periods $\exp 2\pi i F_j$ along the cycles K_{2j} $(j = 1,\ldots,g)$.

Let $a(s) \in L_p^{0,1}(M)$, $p > 2$, be such a coefficient that

$$\frac{1}{\pi} \iint_{\hat{M}} a(p) w_j'(p)\, d\sigma_p = F_j , \quad j = 1,\ldots,g . \tag{9.36}$$

Then the function

$$u_0(s) = \exp \frac{1}{\pi} \iint_{\hat{M}} a(p) m(p,s)\, d\sigma_p \tag{9.37}$$

is a regular solution of the equation

$$\bar{\partial} u + a(s) u = 0 \tag{9.38}$$

without zeros and poles having the multiplicative periods $\exp 2\pi i F_j$ along the cycles K_{2j} $(j = 1,\ldots,g)$.

Then the function $u(s) = u_0(s) \exp\{-\omega(s)\}$ is a single-valued solution of the equation (9.38) which is a multiple of the divisor $\gamma = \kappa q_0$. Depending on the existence of such solutions, the Jacobi inversion problem is solvable or unsolvable.

By the Riemann-Roch theorem the dimension of the space of such functions satisfies the inequality

$$\dim L(\kappa q_0) \geq 2\kappa - 2g + 1$$

and, consequently, the problem is solvable if $\kappa \geq g$.

Note. The connection between the Jacobi inversion problem and the Riemann-Roch theorem is ascertained in passing from analytic functions to solutions of the equation (9.38). Instead of equation (9.38) one can use the analytic solution of the Riemann problem with appropriate periods,

$$\Phi_0(s) = \exp \frac{1}{2\pi i} \int_L \ln G(p) m(p,s) dz(p) ,$$

$$\int_L \ln G \, dw_j = \frac{1}{2\pi i} F_j , \quad j = 1,\dots,g . \tag{9.39}$$

The equivalence of equation (9.38) and the Riemann problem is fundamental as it will be shown in the following section.

D. The classification of linear complex bundles*)

Let $u(s)$ be a solution of equation (9.10) and $N = \{U_i, i \in J\}$ be an open covering of the surface M (see § 3). If $z_j = z_j(s)$ is a fixed local coordinate in U_j, then we have the local representation

$$u(s) = \varphi_j(z_j) \exp \omega_j(s) , \quad \omega_j(s) = \frac{1}{\pi} \iint_{U_j} a(z_j(p)) \frac{d\sigma_{z_j}(p)}{z_j(p) - z_j(s)} . \tag{9.40}$$

We obtain the complex line bundle B_a determined by the cocycle

$$g_{ij}(s) = \frac{\omega_i(s)}{\omega_j(s)} , \quad s \in U_i \cap U_j . \tag{9.41}$$

Conversely, if $B \to M$ is a complex line bundle over M determined by some cocycle $\{h_{ij}\} \in Z^1(\Omega^*)$, then there exists B_a equivalent to the bundle B.

Divide the surface M into the domains T^+ and T^- by a contour L. In the domains $T^\pm$ the bundles $B \to T^+$ and $B \to T^-$, respectively are trivial (see H. Cartan [a], J.P. Serre [a], O. Forster [a]). Therefore, in these domains the transition functions are representable in the form $h_{ij} = h_i h_j^{-1}$ where the h_i are holomorphic functions different from zero. Let the domain U_i be divided by the contour L into the domains $U_i^\pm$. Then in $U_i^\pm$ one can define the cochains $\{h_i^\pm\}$ and

$$h_{ij} = h_i^+(h_j^+)^{-1} = h_i^-(h_j^-)^{-1} \quad \text{on} \quad L \cap U_j \cap U_i .$$

---

*) For simplicity we limit ourselves to the case of the line bundles. The content of this section is valid for complex vector bundles (see H. Röhrl [a], A. Grothendieck [a]) .

Therefore,

$$\frac{h_i^+(p)}{h_i^-(p)} = \frac{h_j^+(p)}{h_j^-(p)} \text{ def} = G(p) \quad \text{on } L \tag{9.42}$$

and, consequently the cochain $\{h_i\}$ determines a germ of the solution of the Riemann problem

$$F^+ = GF^- . \tag{9.43}$$

Conversely, the sheaf of germs of the solutions $\{\varphi_i^\pm\}$ of the problem (9.43) determines the bundle B_G with transition functions

$$h_{ij} = \varphi_i/\varphi_j .$$

This relation between complex bundles and the Riemann problem was established by A. Grothendieck [a] and H. Röhrl [a] (see also Yu.L. Rodin [p]). The connection between the Riemann problem and the system (9.10) is established easily (B.V. Bojarskii [a], H. Röhrl [a], Yu.L. Rodin [d]).

In the domains $T^\pm$ solutions of the system (9.10) are representable in the form

$$u^\pm(s) = \varphi^\pm(s) \exp \frac{1}{\pi} \iint_{T^\pm} a(p) m_\delta^\pm(p,s)\, d\sigma_p . \tag{9.44}$$

The kernels $m_\delta^\pm(p,s)$ (see (5.25)) are chosen such that the characteristic divisors $\delta^\pm$ are located in the domains $T^\pm$.

We obtain the relation

$$\varphi^+(s) = G(s)\varphi^-(s) , \quad s \in L , \tag{9.45}$$

where

$$G(s) = \exp \left\{- \frac{1}{\pi} \iint_{T^+} a(p) m_\delta^+(p,s)\, d\sigma_p + \frac{1}{\pi} \iint_{T^-} a(p) m_\delta^-(p,s)\, d\sigma_p\right\} .$$

Conversely, let $F(s)$ be an analytic solution of the Riemann problem

$$F^+(p) = G(p)F^-(p) , \quad p \in L .$$

Continue the function $G(p)$ into the domain T^- such that $G(s) \in D_p(T^-)$, $\bar{\partial} G \in L_p(T^-)$, $p > 2$. Then the function

$$u(s) = \begin{cases} F^+(s), & s \in T^+, \\ G(s)F^-(s), & s \in T^-, \end{cases} \tag{9.46}$$

is a solution of (9.10) where

$$a(s) = \begin{cases} 0, & s \in T^+, \\ -G^{-1}(s)\bar{\partial} G, & s \in T^-. \end{cases}$$

E. Boundary value problems for generalized analytic functions

1. Here we consider the Riemann problem

$$u^+(p) = G(p)u^-(p) + g(p) \qquad p \in L \tag{9.47}$$

for the equation

$$\underset{\sim}{\partial} u \equiv \bar{\partial} u + a(s)u(s) + b(s)\overline{u(s)} = 0 \tag{9.48}$$

and the conjugate problem for covariants of the type (1.0)

$$v^+(p) = \frac{1}{G(p)} v^-(p), \qquad p \in L \tag{9.49}$$

for the equation

$$\underset{\sim}{\partial}^* v \equiv -\bar{\partial} v + a(s)v(s) + \overline{b(s)}\,\overline{v(s)} = 0. \tag{9.50}$$

The assumptions on the contour, the coefficients of the equation and the boundary functions are the same as above.

Theorem 9.1. The numbers of linear independent solutions l of the homogeneous problem (9.47) and h of the problem (9.49) are connected by the relation

$$l - h = 2\kappa - 2g + 2. \tag{9.51}$$

For the solvability of the problem (9.47) it is necessary and sufficient that

$$\operatorname{Re} \frac{1}{2i} \int_L g(p) v_j^+(p) \, dz(p) = 0 , \qquad j = 1,\dots,h \tag{9.52}$$

where v_j $(j = 1,\dots,h)$ is a complete system of solutions of the homogeneous problem (9.49).

Let $f^{\pm}(s)$ be an analytic solution of the problem

$$f^+(p) = G(p) f^-(p) \tag{9.53}$$

determined by the divisor (f), $\deg (f) = \kappa = \operatorname{ind}_L G$. As it was shown above, such solutions exist if they are allowed to have $g - \kappa$ poles. If u and v are solutions of the problems (9.47) and (9.49), respectively then

$$u_1(s) = \frac{u(s)}{f(s)} , \qquad v_1(s) = v(s) f(s) \tag{9.54}$$

are the function and the covariant belonging to the spaces $L(f)$ and $H(f)$ of functions and covariants, respectively which are multiples of the divisor $-(f)$ and the divisor (f), respectively for the equations

$$\bar{\partial} u_1 + a_1 u_1 + b_1 \overline{u_1} = 0 ,$$

$$-\bar{\partial} v_1 + a_1 v_1 + \overline{b_1 v_1} = 0 , \tag{9.55}$$

$$a_1(s) = a(s) , \; b_1(s) = b(s) \frac{\overline{f(s)}}{f(s)} .$$

The relation (9.51) is the Riemann-Roch theorem for the equations (9.55).

In order to study the inhomogeneous problem (9.47) we prove two lemmas.

Lemma 1. For the solvability of the Plemelj-Sokhotskii problem

$$u^+ - u^- = g \quad \text{on } L \tag{9.56}$$

in the class of regular solutions of the equation (9.48) it is necessary and sufficient that

$$\operatorname{Re} \frac{1}{2i} \int_L g(p) v_j(p) dz(p) = 0 \;, \qquad j = 1,\dots,h_0 \tag{9.57}$$

where v_j $(j = 1,\dots,h_0)$ is a complete set of covariants of the first kind for equation (9.50).

The necessity of these conditions is verified directly by the Green formula. To show that they are sufficient we represent the function $g(s)$ in the form

$$g(p) = u_0^+(p) - u_0^-(p) \qquad \text{on } L \tag{9.58}$$

where $u_0^{\pm}(s)$ in $T^{\pm}$ are regular single-valued solutions of the inhomogeneous Cauchy-Riemann equation

$$\bar{\partial} u_0 + a_0(s) = 0 \qquad s \in T^{\pm}$$

with a proper coefficient a_0 (see (9.8)). If a solution of the problem (9.56) exists, the function

$$U(s) = u(s) - u_0(s)$$

is on M a regular solution of the equation

$$\bar{\partial} U + aU + b\bar{U} = a_0 - au_0 - b\bar{u}_0 \;. \tag{9.59}$$

From equation (9.57) we obtain

$$0 = \operatorname{Im} \int_L g v_j \, dz(p) = \frac{1}{2i} \int_L g v_j dz - \frac{1}{2i} \int_L \overline{g v_j dz} =$$

$$= \frac{1}{2i} \int_L (u_0^+ - u_0^-) \, v_j \, dz - \frac{1}{2i} \int_L (\bar{u}_0^+ - \bar{u}_0^-) \, \bar{v}_j d\bar{z} =$$

$$= \iint_M \bar{\partial}(u_0 v_j) \, d\sigma_p + \iint_M \overline{\bar{\partial}(u_0 v_j)} \, d\sigma_p = \tag{9.60}$$

$$= 2 \operatorname{Re} \iint_M [-a_0(p) v_j(p) + a(p) u_0(p) v_j(p) + \overline{b(p)} u_0(p) \overline{v_j(p)}\} \, d\sigma_p =$$

$$= 2 \operatorname{Re} \iint_M [-a_0(p) + a(p) u_0(p) + b(p) \overline{u_0(p)}] \, v_j(p) \, d\sigma_p \;, \quad j = 1,\dots,h_0 .$$

Therefore, the equation (9.59) satisfies the conditions (4.19) and hence the problem (9.56) is solvable.

Lemma 2. In order that problem (9.56) is solvable in the class of solutions of the equation (9.48) which are multiples of the divisor $-\gamma$ it is necessary and sufficient that the conditions (9.57) are valid. Here v_j $(j = 1,\dots,h)$ is a basis of the space $H(\gamma)$ of covariants for the equation (9.50) which are multiples of the divisor γ .

The proof coincides with the previous one if the operators $\underset{\sim}{\partial}$ and $\underset{\sim}{\partial}^*$ are replaced by the operators $\underset{\sim}{\partial}_\gamma$ and $\underset{\sim}{\partial}^*_\gamma$, respectively in the bundle B_γ and equation (4.19') is used instead of (4.19).

We now pass to prove the second part of Theorem 9.1.

The function $u_1(s)$ from (9.54) satisfies the boundary condition

$$u_1^+(p) - u_1^-(p) = \frac{g(p)}{f^+(p)}, \qquad p \in L \tag{9.61}$$

in the class of solutions of equation (9.55) which are multiples of the divisor $-(f)$. The statement (9.52) follows directly from Lemma 2.

Note. There are two other approaches to study the Riemann problem for generalized analytic functions without using algebraic topological methods.

Theorem 9.1 can be proved directly by passing from the differential equation $\underset{\sim}{\partial}u = F$ to the integral equation $Su = TF + Z$.

Another approach consists in the construction of the analog of the kernel of the Cauchy type integral for generalized analytic functions and reducing the Riemann problem to a singular integral equation. On the complex plane this was done by I.N. Vekua [a] for the Riemann-Hilbert problem. On a Riemann surface of nonzero genus this approach is connected with much trouble but it can be realized completely (Yu.L. Rodin [e], [f], see also W. Koppelman [a]).

2. Let T be a domain on a closed Riemann surface M bounded by a Ljapounov contour $L = \sum_{j=0}^{m} L_j$ consisting of $m + 1$ curves without self-intersections, and let $\alpha(p)$, $\beta(p)$, $\gamma(p)$ be H-continuous functions on L .

The Riemann-Hilbert problem: find a solution of equation (9.10) regular in T and satisfying the boundary condition

$$\text{Re}\,\{[\alpha(p) - i\beta(t)]u(p)\} = \gamma(p) \quad \text{on} \quad L\,. \tag{9.62}$$

The simplest approach to this problem is the following one (N.I. Muskhelishvili [a]; for Riemann surfaces Yu.L. Rodin [a,e]).

Let $\tilde{T}$ be a second copy of the domain T where all local coordinates $z(p)$ are replaced by complex-conjugate ones, $\overline{z(p)}$. If we confine ourselves to the class of local coordinates which are real on L then the curves L and $\tilde{L} = \partial\tilde{T}$ can be identified. The obtained Riemann surface $D = T + \tilde{T} + L$ of genus $2g + m$ (g is the genus of T) is called the double of the surface T.

Continuing the function $u(s)$ by symmetry

$$\begin{aligned} u^+(s) &= u(s)\,, \quad s \in T\,, \\ u^-(s) &= \overline{u(\bar{s})}\,, \quad s \in \tilde{T}\,, \end{aligned} \tag{9.63}$$

from (9.62) it follows the Riemann boundary condition

$$(\alpha - i\beta)u^+(p) + (\alpha + i\beta)u^-(p) = 2\gamma(p)\,, \quad p \in L\,. \tag{9.64}$$

<u>Theorem 9.2</u>. The difference between the number of solutions of the homogeneous problem (9.62) and the number h of solutions of the conjugate problem

$$\text{Re}\; i(\alpha + i\beta)v(p) = 0 \tag{9.65}$$

for the equation $\underset{\sim}{\partial}{}^*v = 0$ for covariants of the type (1,0) is equal to

$$1 - h = 2\,\text{ind}_L\,(\alpha + i\beta) - 4g - 2m + 2\,. \tag{9.66}$$

For the solvability of the inhomogeneous problem it is necessary and sufficient that

$$\text{Re}\,\frac{1}{2i}\int_L \gamma(p)v_j(p)\,dz(p) = 0\,, \quad j = 1,\dots,h \tag{9.67}$$

where v_j $(j = 1,\dots,h)$ is a complete system of solutions of problem (9.65).

CHAPTER 4

NONLINEAR ASPECTS OF GENERALIZED ANALYTIC FUNCTION THEORY

§ 10. Multiplicative multi-valued solutions

A. Multiplicative constants. Existence

Let $u(s)$ be a generalized solution of the equation

$$\underset{\sim}{\partial} u = \bar{\partial} u + au + b\bar{u} = 0 \quad , \; a,b \in L_r^{0,1}(M) \quad , \; r > 2 \; . \tag{10.1}$$

It may be represented in the form

$$u(s) = \varphi(s) \exp \frac{1}{\pi} \iint\limits_{\hat{M}} [a(p) + b(p) \frac{\overline{u(p)}}{u(p)}] \, M(p,s) \, d\sigma_p \tag{10.2}$$

where $\varphi(s)$ is analytic on M (see (1.31)). So long as the kernel of this representation is multi-valued, the function $\varphi(s)$ in general has multiplicative periods along the cycles K_j $(j = 1,\ldots,2g)$. Conversely, substituting the expression $\exp w(s)$ instead of $\varphi(s)$, where $w(s)$ is an Abelian integral of the first or third kind, we consider (10.2) as a nonlinear integral equation and obtain a multi-valued solution on M with multiplicative periods. Fix a branch of the kernel by the condition (5.17). Then we obtain solutions on the cutted surface $\hat{M}$ corresponding to the chosen branch of the kernel. This branch of the solution satisfies equation (10.1). If all periods of a solution are equal to one, we obtain a single-valued solution of the equation $\underset{\sim}{\partial} u = 0$.

In particular, if $\varphi(s) \equiv$ const., the equation

$$u(s) = c \exp \frac{1}{\pi} \iint\limits_{\hat{M}} [a(p) + b(p) \frac{\overline{u(p)}}{u(p)}] \, M(p,s) \, d\sigma_p \tag{10.3}$$

determindes solutions without zeros and poles. We will call these solutions multiplicative constants. If they are single-valued, we obtain the generalized constants (see Chapter 2).

In contrast to generalized constants, multiplicative constants exist always.

Indeed, the operator

$$Ru = T(a + b\frac{\bar{u}}{u}) = \frac{1}{\pi}\iint_{\hat{M}} [a(p) + b(p)\frac{\overline{u(p)}}{u(p)}]\, M(p,s)\, d\sigma_p \qquad (10.4)$$

is compact in the space $L_q(\hat{M})$, $\frac{1}{2} - \frac{1}{r} < \frac{1}{q} < 1$ and maps this space into itself. The estimation

$$\| Ru \|_{L_q(\hat{M})} \le M_{r,q} \left(\| a \|_{L_r(\hat{M})} + \| b \|_{L_r(\hat{M})}\right) \qquad (10.5)$$

is valid. Here $M_{r,q}$ is some constant. The proof is the same as that for Theorem 6.2.

From (10.5) it follows that a sphere of radius

$$\rho \ge M_{r,q} \left(\| a \|_{L_r(\hat{M})} + \| b \|_{L_r(\hat{M})}\right)$$

is mapped by the operator R into itself. By the Shauder principle the operator $c \exp Ru$ has a fixed point. As it was shown above, it is a multiplicative constant. Therefore, we obtain

<u>Theorem 10.1</u> (Yu.L. Rodin [g]). The equation (10.3) is solvable for any constant c in the space $L_q(M)$, $\frac{1}{2} - \frac{1}{r} < \frac{1}{q} < 1$. Its solution is a multiplicative constant satisfying the condition $u(q_0) = c$.

B. Multiplicative constants. Uniqueness

<u>Theorem 10.2</u> (Yu.L. Rodin [g,h]). The fixed point of the operator $\exp R$ is unique in any space $L_q(M)$, $\frac{1}{2} - \frac{1}{r} < \frac{1}{q} < 1$.

Let $h(q) = -2 \arg u(q)$. Taking the argument of both sides of the equation $u = \exp Ru$ we get

$$h(s) = Kh = \iint_{\hat{M}} [A(p,s) - B(p,s) \sin [h(p) + C(p,s)]]\, d\sigma_p , \qquad (10.6)$$

$$A(p,s) = -\frac{2}{\pi} \operatorname{Im} [a(p)M(p,s)]$$

$$B(p,s) = \frac{2}{\pi} |b(p)M(p,s)| ,$$

$$C(p,s) = \arg [b(p)M(p,s)] .$$

Solving equation (10.6) we get

$$u(q) = |\exp T(a + b \exp ih)| .$$

Let $h_0 = -2 \arg u_0(q)$ be a fixed point of the operator K. The Fréchet derivative of this operator at the point h_0 is equal to

$$B\psi = - \iint_{\hat{M}} B(p,s)\cos [h_0(p) + C(p,s)]\psi(p)\, d\sigma_p . \tag{10.7}$$

Investigate the spectrum of B. Introduce the function $\Psi(q)$, $\operatorname{Im} \Psi = \psi$, $\operatorname{Re} \Psi(q) = \operatorname{Re} \frac{2i}{\pi} \iint_{\hat{M}} b_0(p)\psi(p)M(p,q)\, d\sigma_p$. Then the equation $B\psi = \lambda\psi$ can be written into the form

$$\Psi(q) - \frac{1}{\pi} \iint_{\hat{M}} b_0(p)\, [\Psi(p) - \overline{\Psi}(p)]M(p,q)\, d\sigma_p = 0 , \tag{10.8}$$

$$b_0(p) = - \frac{1}{\lambda}\, b(p) \frac{\overline{u_0(p)}}{u_0(p)} .$$

<u>Lemma</u>. Equation (10.8) has no non-trivial solutions for $b_0 \in L_r^{0,1}$, $r > 2$.

We shall show that the adjoint equation

$$v(q) - \frac{1}{\pi} \iint_{\hat{M}} [b_0(p)v(p) - \overline{b_0(p)}\,\overline{v(p)}]M(q,p)\, d\sigma_p = 0 \tag{10.9}$$

has no non-trivial solutions. If $v(q)$ is a solution of equation (10.9), then

$$\bar{\partial} v + b_0 v - \overline{b_0 v} = 0 . \tag{10.10}$$

In this case the form $i\omega = v\, dz(q) - \bar{v}\, \overline{dz(q)}$ is closed, i.e. $-\omega = df(q)$. It means that $v = -i\partial f$ and $\bar{v} = i\bar{\partial} f$. Therefore, the value $f(q)$ is real and satisfies the equation

$$\frac{\partial^2 f}{\partial z \partial \bar{z}} + b_0 \frac{\partial f}{\partial z} + \bar{b}_0 \frac{\partial f}{\partial \bar{z}} = 0 . \tag{10.11}$$

Since $v(q)$ may have a single singularity at the point $p = s_0$, the value $f(q)$ has a finite upper or lower bond on M. Rewrite equation (10.9) into the form

$$i\partial_q f(q) - \frac{1}{\pi}\iint_{\hat{M}} \bar{\partial}_p v(p)\, \partial_q [\Omega_{s_0q}(p) - \Omega_{s_0q}(q_0)]\, d\sigma_p = 0 \qquad (10.12)$$

and suppose

$$i\,\overline{F(q)} = i\,f(q) - \frac{1}{\pi}\iint_{\hat{M}} \bar{\partial}_p v\, [\Omega_{s_0q}(p) - \Omega_{s_0q}(q_0)]\, d\sigma_p \,. \qquad (10.13)$$

The function $F(q)$ is analytic on M. The second summand of the right hand side of (10.13) is single-valued on M. Indeed, by (5.13), (5.2), and (5.7)

$$\int_{K_j} d_q [\Omega_{s_0q}(p) - \Omega_{s_0q}(q_0)] = \int_{K_j} d\Omega_{q_0p}(q) -$$

$$- 2\pi i \sum_{l=1}^{g} \{\operatorname{Im}\int_{q_0}^{p} d\theta_{2l-1}\ \operatorname{Im}\int_{K_j} d\theta_{2l} - \operatorname{Im}\int_{q_0}^{p} d\theta_{2l}\ \operatorname{Im}\int_{K_j} d\theta_{2l-1}\} = 0,$$

$$j = 1,\dots,2g \,. \qquad (10.14)$$

Therefore, periods of $\overline{F(q)}$ coincide with periods of $f(q)$ and hence they are real. This means that $F \equiv$ const.. Whence it follows that $f(q)$ is single-valued. By the maximum principle for equation (10.11) (L. Bers, L. Nirenberg [a,b] ; L. Bers, F. John, M. Schechter [a]) $f \equiv$ const. and hence $v \equiv 0$.

We will briefly show how one can obtain the statement of the theorem. For details see M.A. Krasnoselskii [a], Chapter 2,3.

The point h_0 is an isolated fixed point of the operator K. Indeed, let S_ε be a sphere of radius $\varepsilon > 0$ with centre at h_0 in $L_q(M)$, $\frac{1}{2} - \frac{1}{r} < \frac{1}{q} < 1$, and $h_0 + g$ be a point of the sphere. Then

$$\| K(h_0 + g) - (h_0 + g) \| = \| K(h_0 + g) - Kh_0 - g \| =$$

$$= \| Bg - g + \alpha g \| \geq \| Bg - g \| - \| \alpha g \| \geq \varepsilon_1 \| g \|$$

since 1 is no eigenvalue of the operator B. This means that the operator K has no fixed points on S_ε for small ε. The operator $E - K$ is homotopic to $E - B$ on S_ε. Indeed, the vector field

$$\Phi(g,t) = t\,[(h_0 + g) - K(h_0 + g)] + (1 - t)\,[(h_0 + g) - (h_0 + Bg)]$$

has no zeros on S_ε and $\Phi(g,0) = g - Bg$ and $\Phi(g,1) = (h_0 + g) - K(h_0 + g)$. Therefore, the rotations of the operators K and B on S_ε coincide. The rotation of K on S_ε is equal to the index of the point h_0 ; the rotation of B on S_ε is equal to $(-1)^\beta$ where β is the sum of multiplicities of all eigenvalues of the operator B belonging to $[1,\infty)$. As it has been shown above, $\beta = 0$ and hence the index of any fixed point of the operator K is equal to $+1$. Since the operator K is compact, its fixed points form a compact set. Since this set is discret, it is finete. Therefore, the operator K has a finite number of fixed points. Because the operator K maps the sphere S_R of a large radius R into itself, this operator is homotopic to the identity operator E on S_R . Hence the rotation of K on S_R is equal to $+1$. On the other hand, the rotation of K on S_R is equal to the sum of the indices of the fixed points. Whence it follows that there is only a single fixed point of the operator K .

C. Abel's theorem

The Abel's problem of the existence of a solution to the equation $\underset{\sim}{\partial} u = 0$ determined by the divisor $\gamma = \sum_{k=1}^{n} \alpha_k p_k$, $\deg \gamma = 0$ is reducable to the equation

$$u(q) = \exp\left\{\sum_{k=1}^{n} \alpha_k \Omega_{p_0 p_k}(q) + \sum_{j=1}^{g} c_j w_j(q)\right\} Ru \tag{10.15}$$

where $\{w_j\}_1^g$ is a basis of the first kind Abelian integrals and $\Omega_{p_0 p_k}$ are the third kind Abelian integrals. Let

$$u(q) = v(q)H(q)\ , \ H(q) = \exp\left\{\sum_{k=1}^{n} \alpha_k \Omega_{p_0 p_k}(q) + \sum_{j=1}^{g} c_j w_j(q)\right\} . \tag{10.16}$$

We obtain the equation $v = \tilde{R}v$. The operator $\tilde{R}$ is obtained from the operator R by the exchange of the coefficient $b(q)$ by $b(q)\overline{H(q)}/H(q)$. Let $v(c_1,\ldots,c_g)$ be a fixed point of the operator $\tilde{R}$ with multiplicative periods $v_j(c_1,\ldots,c_g)$ along the cycles

K_j ($j = 1,\ldots,2g$) . Thus, the Abel's problem is reduced to the system of equations

$$v_j(c_1,\ldots,c_g) = 1 , \qquad j = 1,\ldots,2g . \tag{10.17}$$

CHAPTER 5

SOME GENERALIZATIONS AND APPLICATIONS

§ 11. Singular cases

We consider here the case where the coefficients of the operator $\underset{\sim}{\partial}$ have isoleted singularities of first order. This situation was investigated for case $g = 0$ by L.C. Mikhailov [a]. He pointed out several patalogies (infringements of the argument principle, the Liouville theorem and so on). A. Turakulov [a,b] established that these phenomena are consequences of the fact that in this case the index of the operator $\underset{\sim}{\partial}$ depends on the surface's genus as well as on the divisor's degree and the so-called $a(q)$ coefficient's defect. It appears that the coefficient $b(q)$ influences the index, too.

Further we briefly will state the K.L. Volkoviskii paper [a] on generalized analytic functions on surfaces of infinite genus.

1. Generalized analytic functions with singular coefficients

A. Plane examples. Integral operators

We begin with simple examples. Consider the equation

$$\bar{\partial}u + \frac{a}{\bar{z}}\, u = 0 \;, \qquad a = \text{const.}, \tag{11.1}$$

in the disk $|z| < 1$. We have

$$u(z) = |z|^{-2a} f(z) \tag{11.2}$$

where $f(z)$ is analytic. A solution of the equation

$$\bar{\partial}u + \frac{a}{z}\, u = 0 \tag{11.3}$$

is represented by the formula

$$u(z) = \exp\left\{-\frac{a\bar{z}}{z}\right\} f(z) \tag{11.4}$$

where $f(z)$ is analytic. The solutions (11.2) and (11.4) are different in principle. The function (11.4) is discontinuous along a line determining some single-valued branch (for example, along the real semi-axes); the function (11.2) has a zero or a pole at $z = 0$ disturbing the equality of the numbers of zeros and poles of u and f. It appears that this phenomenon is characteristic only for coefficients with poles of the type $1/\bar{z}$. Note that the function (11.2) for $f(z) \equiv 1$ is a solution of the equation

$$\bar{\partial} u + \frac{a}{\bar{z}} \bar{u} = 0 . \tag{11.5}$$

Therefore, the phenomena related to a breach of the argument principle are possible for equations of the type (11.5).

Let $p_1, \ldots, p_{n+1}$ be some points of the surface M, $d = \bigcup_{j=1}^{n+1} \{p_j\}$, and $D \subset M$ be some simply-connected coordinate domain, such that $d \subset D$. We consider the operator

$$\underset{\sim}{\partial} u \equiv \bar{\partial} u + au + b\bar{u} \tag{11.6}$$

the coefficients of which in the domain D are represented in the form

$$a(z) = \sum_{j=1}^{n+1} \frac{a_j(z)}{|z-z_j|} + a_0(z) , \qquad z = z(p) , \ p \in D , \tag{11.7}$$

$$b(z) = \sum_{j=1}^{n+1} \frac{b_j(z)}{|z-z_j|} + b_0(z) , \tag{11.8}$$

$$z_j = z(p_j) , \qquad j = 1, \ldots, n+1 . \tag{11.9}$$

Here $z = z(p)$ is some fixed local coordinate in the domain D and $a_j(z)$, $b_j(z)$ are bounded measurable functions. In the domain $M \setminus D$ we suppose $a, b \in L_p^{0,1}(M \setminus D)$, $p > 2$.

Let $\alpha = (\alpha_1, \ldots, \alpha_{n+1})$ be a multi-index such that $0 < \alpha < 1$ (i.e. $0 < \alpha_j < 1$, $j = 1, \ldots, n+1$). Determine the piecewise continuous function

$$d^{\alpha}(q) = \begin{cases} \prod_{j=1}^{n+1} |z - z_j|^{\alpha_j} , & q \in D , \\ 1 , & q \in M \setminus D , \end{cases} \qquad (z = z(q)) \tag{11.10}$$

and consider the space $S(d,\alpha,M)$ of functions $f(q)$ such that $d^{\alpha}(q)f(q)$ is a bounded measurable function with norm

$$\| f \|_{S(d,\alpha,M)} = \text{vrai max } \{d^{\alpha}(q)|f(q)|\} . \tag{11.11}$$

Consider the operator

$$Pu = -\frac{1}{\pi} \iint_{\hat{M}} [a(q)u(q) + b(q)\overline{u(q)}]m(q,s)\, d\sigma_q \tag{11.12}$$

in the space $S(d,\alpha,M)$.

Theorem 11.1. The operator

$$P: S(d,\alpha,M) \to S(d,\alpha,M) , \qquad 0 < \alpha < 1 ,$$

is bounded. The function $g(s) = (Pu)(s)$ has generalized derivatives and

$$\bar{\partial} g = a(s)u(s) + b(s)\overline{u(s)} . \tag{11.13}$$

The operator P is representable in the form

$$P = P_0 + P_1 ,$$

where

$$P_0 u = -\frac{1}{\pi} \iint_{z(D)} \{[a(t) - a_0(t)]u(t) + [b(t) - b_0(t)]\overline{u(t)}\} \frac{d\sigma_t}{t-z} \tag{11.14}$$

$$t = z(q) , \qquad z = z(s)$$

and $P_1 = P - P_0$ is compact in the space $S(d,\alpha,D)$.
In D we have the estimation

$$\| P_0 u \|_{S(d,\alpha,D)} \le \frac{1}{\pi} \text{vrai max} \prod_{j=1}^{n+1} |z - z_j|^{\alpha_j} \cdot$$

$$\cdot \sum_{j=1}^{n+1} \left\{ \iint_{z(D)} \frac{|a_j(t)u(t)|d\sigma_t}{|t-z||t-z_j|} + \iint_{z(D)} \frac{|b_j(t)\overline{u(t)}|d\sigma_t}{|t-z||t-z_j|} \right\} \le$$

$$\leq \frac{1}{\pi} \text{ vrai max} \prod_{j=1}^{n+1} |z-z_j|^{\alpha_j} \sum_{j=1}^{n+1} \iint_{z(D)} \frac{\{|a_j(t)|+|b_j(t)|\}d\sigma_t}{|t-z|\,|t-z_j|^{1+\alpha_j}} \| u \|_{S(d,\alpha,D)} .$$

Because of equation (11.7)

$$\text{vrai max } \{|a_j(t)| + |b_j(t)|\} \leq N , \quad j = 1,\dots,n+1 .$$

Further, we use the well-known estimation (I.N. Vekua [a], chapter 1 (6.7))

$$\iint_{z(D)} \frac{d\sigma_t}{|t-z|\,|t-z|^{1+\alpha_j}} \leq M(\alpha_j)\ |z-z_j|^{-\alpha_j} .$$

We obtain the estimation

$$\| P_0 u \|_{S(d,\alpha,D)} \leq M(\alpha)N \ \| u \|_{S(d,\alpha,D)} . \tag{11.15}$$

The function $au + b\bar{u}$ has singularities at the points p_j the orders of which are not greater than $1 + \alpha_j < 2$ $(j = 1,\dots,n+1)$. Therefore, $au + b\bar{u} \in L_1^{0,1}(M)$ whence equation (11.13) follows by Theorem 1.3.

Consider the space $L_1^{1,0}(d,-\alpha,M)$ of covariants of the type $(1,0)$ with the norm

$$\| v \|_{L_1^{1,0}(d,-\alpha,M)} = \iint_{\hat{M}} d^{-\alpha}(s)\ |a(s)v(s) + \overline{b(s)}\,\overline{v(s)}|\ d\sigma_s . \tag{11.16}$$

The duality of the spaces $S(d,\alpha,M)$ and $L_1^{1,0}(d,-\alpha,M)$ is defined by the bilinear form determined on the product of these spaces

$$(u,v) = \text{Re} \iint_{\hat{M}} u(s)\ [a(s)v(s) + \overline{b(s)}\,\overline{v(s)}]\ d\sigma_s =$$

$$= \text{Re} \iint_{\hat{M}} [a(s)u(s) + b(s)\overline{u(s)}]v(s)\ d\sigma_s . \tag{11.17}$$

The adjoint operator to P has the form

$$P^*v = -\frac{1}{\pi}\iint\limits_{\hat{M}} [a(s)v(s) + \overline{b(s)}\,\overline{v(s)}]m(q,s)\,d\sigma_s . \tag{11.18}$$

The following statement can be easily verified

<u>Theorem 11.2</u>. The operator P^* is bounded in the space $L_1^{1,0}(d,-\alpha,M)$.

<u>Note</u>. The norm (11.16) is equal to zero if $v(q)$ is different from zero only in those domains where $a = b = 0$.
The proof requires a more deep analysis of this construction omitted here.

From Theorem 11.1 and Theorem 11.2 it follows that the operators P and P^* are Noetherian (Φ-operators) (I.C. Gohberg, M.G. Krein [a]). Whence it follows

<u>Theorem 11.3</u>. The equations $Su = u + Pu = 0$ and $S^*v = v + P^*v = 0$ have finite numbers g_0 and g_0' of solutions in the spaces $S(d,\alpha,M)$ and $L_1^{1,0}(d,-\alpha,M)$, respectively. For the solvability of the inhomogeneous equations $Su = f$, $S^*v = h$ it is necessary and sufficient that

$$\begin{aligned} (f,v_j) &= 0 , \quad j = 1,\dots,g_0' , \\ (u_j,h) &= 0 , \quad j = 1,\dots,g_0 , \end{aligned} \tag{11.19}$$

respectively. Here u_j $(j = 1,\dots,g_0)$ and v_j $(j = 1,\dots,g_0')$ are complete systems of solutions of the homogeneous equations $Su = 0$ and $S^*v = 0$.

The index of the operator S

$$\operatorname{ind} S = g_0 - g_0' \tag{11.20}$$

plays a principal role in the Riemann-Roch theorem.

In the following section we shall obtain the Riemann-Roch theorem and in § 11C we shall consider the cases where ind S can be calculated.

B. The Riemann-Roch theorem

Here we confine ourselves to the case of generalized constants and differentials of the first kind, i.e. we consider the Riemann-Roch theorem for zero divisors. The general case may be investigated by the same method.

The function $u(q)$ is called an α-generalized constant for the operator $\underset{\sim}{\partial}$ if it is a regular solution of the equation $\underset{\sim}{\partial} u = 0$ on $M \setminus d$ and belongs to the space $S(d,\alpha,M)$, $0 < \alpha < 1$.

The covariant $v(q)$ is called an α-covariant if it is a regular solution of the equation $\underset{\sim}{\partial}^* v = 0$ on $M \setminus d$ and belongs to the space $L_1^{1,0}(d,-\alpha,M)$.

Therefore, a generalized constant may possess singularities at the points $p_j \in d$ of orders not greater than α_j and a covariant of the first kind may have singularities of orders less than $1 - \alpha_j$ $(j = 1,\dots,n+1)$.

Below we fix some $0 < \alpha < 1$ and omit this symbol.

Theorem 11.4. The space L_0 of generalized constants of the equation $\underset{\sim}{\partial} u = 0$ coincides with the space of single-valued solutions of the equation $Su = u + Pu = c$ where c is a constant. The space H_0 of the first kind covariants of the equation $\underset{\sim}{\partial}^* u = 0$ coincides with the space of solutions of the equation

$$S^*v = v + P^*v = Z'$$

where Z' is an Abelian covariant of the first kind.

Denote by g_1 the number of single-valued solutions of the equation $Su = 0$ in the space $S(d,\alpha,M)$. From Theorem 11.4 it follows that they are generalized constants. We would like to remind that g_0 is the number of solutions of the equation $Su = 0$. Let $u_1,\dots,u_{g_0}$ be a complete system of solutions of this equation and

$$\Sigma = \begin{bmatrix} (u_1,w_1') & \dots & (u_1,w_g') & (u_1,iw_1') & \dots & (u_1,iw_g') \\ \cdot & \cdot & \cdot & \cdot & \cdot & \cdot \\ (u_{g_0},w_1') & \dots & (u_{g_0},w_g') & (u_{g_0},iw_1') & \dots & (u_{g_0},iw_g') \end{bmatrix} \qquad (11.21)$$

where $w_1',\dots,w_g'$ is the basis (5.1) of the space of first kind Abelian covariants. Then

$$\text{rang}\,\Sigma = g_0 - g_1 . \qquad (11.21')$$

Indeed, by equation (6.4) any zero combination of rows of the matrix Σ corresponds to a single-valued solution of the equation $Su = 0$

and conversely.

Consider the space of Abelian covariants of the first kind satisfying the conditions

$$(u_j, Z') = 0 \ , \quad j = 1,\dots,g_0 \ . \tag{11.22}$$

In virtue of (11.21') the dimension of this space is equal to $2g - g_0 + g_1$.

Consider the space of solutions of $S^*v = Z'$ (Z' is any Abelian covariant of the first kind or zero) and the basis of this space $v_1,\dots,v_m$, $m = 2g - g_0 + g_1 + g_0'$,

$$\begin{aligned} S^*v_j &= 0 \ , \ j = 1,\dots,g_0' \ , \\ S^*v_j &= Z_j' \ , \ j = g_0' + 1,\dots,g_0' + 2g - g_0 + g_1 = m \end{aligned} \tag{11.23}$$

where Z_j' $(j = g_0' + 1,\dots,m)$ is some basis of the space (11.22).

Consider the matrix

$$\Sigma_0 = \begin{bmatrix} (1,v_1) & \dots & (1,v_m) \\ & & \\ (i,v_1) & \dots & (i,v_m) \end{bmatrix} . \tag{11.24}$$

Note that $g_0 - g_0' = \text{Ind } S$ and hence the matrix (11.24) has

$$m = 2g + g_1 - (g_0 - g_0') = 2g + g_1 - \text{Ind } S$$

columns.

Any vanishing linear combination of the columns of the matrix (11.24) determines a generalized analytic covariant $v(q)$ satisfying the conditions

$$(1,v) = (i,v) = 0 \ .$$

Whence it follows that $v \in H_0$. (H_0 is the space of generalized analytic covariants of the first kind; see Theorem 11.4). Therefore,

$$h_0 = \dim H_0 = 2g + g_1 - \text{ind } S - \text{rank } \Sigma_0 \ . \tag{11.25}$$

The dimension l_0 of the space L_0 of generalized constants is equal to

$$l_0 = 2 + g_1 - \text{rank}\ \Sigma_0 . \tag{11.26}$$

Indeed, any vanishing linear combination of the rows of the matrix (11.24) corresponds to the system

$$(c,v_j) = 0 , \quad j = 1,\ldots,2g + g_1 - \text{ind}\ S . \tag{11.27}$$

The first g_0' of these conditions mean that the equation $Su = c$ is solvable (see (11.23)). They show that there exists a singel-valued solution of this equation. The rest of the conditions may be transformed into

$$(c,v_j) = (Su,v_j) = (u,S^*v_j) = (u,Z_j') = 0 , \quad j = g_0' + 1,\ldots,m. \tag{11.28}$$

Let $u_1,\ldots,u_{g_0}$ be a basis of the space of solutions of the equation $Su = 0$. By the equation (6.4), for the existence of a single-valued solution to $Su = c$ it is sufficient that the system

$$\begin{aligned} &\sum_{j=1}^{g_0} c_j\ (u_j,w_k') + (u,w_k') = 0 , \\ &\sum_{j=1}^{g_0} c_j\ (u_j,iw_k') + (u,iw_k') = 0 , \quad k = 1,\ldots,g \end{aligned} \tag{11.29}$$

is solvable. Here u is some solution of the equation $Su = c$. The rank of the system (11.29) coincides with the rank of the augmented matrix Σ. Indeed, let us have a zero combination of the columns of the matrix (u_j,w_k'),

$$(u_j,w') = 0 , \quad j = 1,\ldots,g_0 .$$

It means that the equation $S^*v = w'$ is solvable. Then the covariant w' is representable in the form

$$w' = \sum_{j=g_0'+1}^{m} c_j Z_j' . \tag{11.30}$$

Whence it follows by (11.28) that

$$(u,w') = \sum_{j=g_0'+1}^{m} c_j \, (u,w_j') = 0 \; .$$

Therefore, the space L_0 consists of g_1 single-valued solutions of the equation $Su = 0$ and $2 - \operatorname{rank} \Sigma_0$ single-valued solutions of the equation $Su = c$, c is a constant. This yields (11.26).

Comparing (11.25) with (11.26) we obtain the following result.

Theorem 11.5 (Riemann-Roch).

$$l_0 - h_0 = 2 - 2g + \operatorname{ind} S \; . \tag{11.31}$$

C. The index of the operators S

Firstly we note that for the index of S it is sufficient to calculate the index of the operator $S_0 = E + P_0$ (see (11.14)) since $P - P_0$ is compact.

In general the formula for the index of the operator S_0 is unknown. But we shall point out one interesting case where the index of S_0 can be calculated. Let

$$a(z) = \sum_{j=1}^{n_0} \frac{a_j}{\bar{z}-\bar{z}_j} + \sum_{j=n_0+1}^{n} \frac{a_j(z-z_{n+1})^{\kappa_j}(\bar{z}-\bar{z}_{n+1})^{-1-\kappa_j}}{s_j(z-z_j)^{\kappa_j}(\bar{z}-\bar{z}_j)^{1-\kappa_j}} \, , \tag{11.32}$$

$$1 \le n_0 \le n \; , \; 0 < \kappa_j < 1 \; , \quad z \in G \; ,$$

where a_j $(j = 1,\ldots,n)$ are arbitrary constants and

$$s_j = (z_j - z_{n+1})^{\kappa_j} (\bar{z}_j - \bar{z}_{n+1})^{-1-\kappa_j} \, , \quad j = n_0 + 1,\ldots,n \; . \tag{11.33}$$

A single-valued branch of the function (11.32) is choosen in the domain $G = z(D)$ cutted along curves $\breve{z_j z}_{n+1}$, $\bigcup_{j=1}^{n} \breve{z_j z}_{n+1} = \underline{d}$. The function $b(z)$ is the same as in (11.8) but $b_0(z) \equiv 0$.

Consider the inhomogeneous equation $S_0 u = f$ in the space $S(d,\alpha,D)$ and assume

$$w(z) = u(z) - f(z) \; .$$

Evidently, $w \in S(d,\alpha,D)$. Below this space is denoted by $S(\alpha)$. We obtain the equation

$$S_0 w = g \tag{11.34}$$

where

$$g(z) = \frac{1}{\pi} \iint_G [a(t)f(t) + b(t)\overline{f(t)}] \frac{d\sigma_t}{t-z} . \tag{11.35}$$

Since $af + b\bar{f} \in S(\alpha + 1) \subset L_1$, by Theorems 1.2 and 1.3 $g \in S(\alpha)$ and

$$\bar{\partial} w + aw + b\bar{w} - \bar{\partial} g = 0 . \tag{11.36}$$

From (11.35), (11.36) it follows that the function $w(z)$ is analytically continuable into $\mathbb{C} - G$ and $w(\infty) = 0$ if $a = b = 0$ in $\mathbb{C} - G$. Consider the Riemann boundary problem

$$\varphi^+(t) = = \frac{1}{\Pi(t)} \varphi^-(t) , \quad t \in \partial G , \quad G = z(D) , \tag{11.37}$$

$$\Pi(z) = \begin{cases} \prod_{j=1}^{n_0} |z-z_j|^{2a_j} \prod_{j=n_0+1}^{n} \exp\{\beta_j \left(\frac{z-z_{n+1}}{z-z_j}\right)^{\kappa_j} \left(\frac{\bar{z}-\bar{z}_{n+1}}{\bar{z}-\bar{z}_j}\right)^{-\kappa_j}\}, & z \in G , \\ 1 , & z \in \mathbb{C} \setminus G , \end{cases}$$

$$\beta_j = \frac{a_j}{s_j \kappa_j} \quad \frac{1}{\bar{z}_j - \bar{z}_{n+1}} .$$

The function

$$\ln \Pi(z) = \sum_{j=1}^{n_0} a_j \ln |z-z_j|^2 + \sum_{n_0+1}^{n} \beta_j \left(\frac{z-z_{n+1}}{z-z_j}\right)^{\kappa_j} \left(\frac{\bar{z}-\bar{z}_{n+1}}{\bar{z}-\bar{z}_j}\right)^{-\kappa_j}$$

is single-valued on the contour $\Gamma = \partial G$ and hence the index of problem (11.37) is zero. Its holomorphic solution which is one at infinity has the form

$$\varphi^{\pm}(z) = \exp \{- \frac{1}{2\pi i} \int_\Gamma \frac{\ln \Pi(t)}{t-z} dt\} .$$

In this case the function

$$\omega(z) = M(z)\,\Pi(z)\varphi^{\pm}(z) \quad , \; M(z) = \prod_{j=1}^{n_0} (z-z_j)^{-m_j}$$

$$m_j - 1 \le 2\,\mathrm{Re}\,a_j < m_j \; , \; \mu_j = 2\,\mathrm{Re}\,a_j - m_j + 1 \; , \quad j = 1,\dots,n_0 \tag{11.38}$$

is a solution of the equation

$$\bar{\partial}\omega + a\omega = 0$$

continuable analytically into $\mathbb{C} - G$. It has singularities of the order

$$1 - \mu_j = m_j - 2\,\mathrm{Re}\,a_j < 1$$

at the points $z_1,\dots,z_{n_0}$ and of the order $-m$, $m = \sum_{j=1}^{n_0} m_j$ at infinity. On $\underline{d}$ it may have a discontinuity of the first kind.

We are looking for the function $w(z)$ in the form

$$w(z) = \omega(z)\nu(z) \; , \tag{11.39}$$

$\nu \in S(\beta)$, $\beta = \alpha + \mu - 1$. We choose α such that $0 < \alpha < 1$, $1 < \alpha + \mu < 2$, i.e. $0 < \beta < 1$. The order of the function $\nu(z)$ at infinity is not greater than $m - 1$,

$$\bar{\partial}\nu + b\,\frac{\bar{\omega}}{\omega}\,\bar{\nu} + \frac{af+b\bar{f}}{\omega} = 0 \; . \tag{11.40}$$

We have the integral equation

$$S_1\nu = (E + P_1)\nu \equiv \nu(z) - \frac{1}{\pi}\iint_G b(t)\,\frac{\overline{\omega(t)}}{\omega(t)}\,\overline{\nu(t)}\,\frac{d\sigma_t}{t-z} =$$

$$= \psi(z) + \frac{1}{\pi}\iint_G \frac{a(t)f(t)+b(t)\overline{f(t)}}{\omega(t)}\,\frac{d\sigma_t}{t-z} \tag{11.41}$$

where $\psi(z)$ is analytic. Three terms of equation (11.41) belong to $S(\beta)$. Hence $\psi(z) \in S(\beta)$ too, i.e. the function $\psi(z)$ is holomorphic in the domain G . Since both integrals in (11.41) are analytic at infinity and equal to zero there, $\psi(z)$ has the same order at infinity as $\nu(z)$, i.e.

$$\psi(z) = \begin{cases} P_{m-1}(z) , & m > 0 , \\ 0 , & m \leq 0 , \end{cases} \tag{11.42}$$

where $P_{m-1}(z)$ is a polynom of degree at most $m - 1$.

For the operator S_1 one can obtain the estimation

$$\| P_1 \|_{S(\beta)} \leq b_0 M(\beta) \tag{11.43}$$

where $M(\beta)$ is some constant and

$$b_0 = \sum_j \text{vrai max } |b_j(z)| \tag{11.44}$$

(see (11.8)). Let

$$b_0 < \frac{1}{M(\beta)} . \tag{11.45}$$

In this case the norm of P_1 is less then one and the equation (11.41) can be solved by iteration. At first the condition (11.45) was used by L.G. Mikhailov [a] and we call (11.45) the Mikhailov condition. If the Mikhailov condition is valid, the equation (11.41) has $2m + 1$ solutions for $m \geq 0$. For $m < 0$ we obtain $2|m|$ real conditions for the solvability.

Theorem 11.6. If the Mikhailov condition (11.45) is valid, the index of the operator S is equal to $2m$, $m = \Sigma m_j$ (see (11.38)).

Indeed, the index of the operator S coincides with the index of the operator S_0 since $S - S_0$ is a compact operator.

As it was established above, the number of solutions of the equation $S_0 u = 0$ is equal to

$$l = \begin{cases} 2m , & m \geq 0 \\ 0 , & m < 0 \end{cases} \tag{11.46}$$

(the equation (11.41) has $2m + 1$ solutions for $\psi \not\equiv 0$) . The operator $\tilde{S}_0^*$ in the space $L_1^{1,0}(d,-\alpha,M)$ is determined by the formula

$$\widetilde{S}_0^* v = \begin{cases} -\frac{1}{\pi} \iint_G \{[a(t) - a_0(t)]v(t) + \overline{b(t)}\ \overline{v(t)}\} \frac{d\sigma_t}{t-z}, & p(z) \in D, \\ v(p), & p \in M \setminus D. \end{cases} \tag{11.47}$$

Therefore, the number of solutions of $\widetilde{S}_0^* v = 0$ in $L_1^{1,0}(d,-\alpha,M)$ is equal to the number l' of solution of the equation $S^* v = 0$ in the space $L_1^{1,0}(d,-\alpha,D)$. Since $(a-a_0)v + \overline{bv} \in L_1(D)$, a solution of the equation $S_0^* v = 0$ belongs to $S(d,1-\alpha,D)$. Repeating the reasoning for the operator S_0, one can obtain the relation for the operator S_0^*

$$l' = \begin{cases} 0, & m \geq 0 \\ -2m, & m < 0. \end{cases} \tag{11.48}$$

From (11.46) and (11.48) the statement of the theorem follows.

2. Open surfaces

An investigation of surfaces of infinite genus is related with several principal difficulties because the index of the Cauchy-Riemann equation is infinite: $l_0 = 2$ (1 and i) and $h_0 = \infty$. The approach to the Riemann-Roch theorem was proposed by Yu.L. Rodin [i] and then was developed by K.L. Volkoviskii [a]. For further results for the analytic case see Yu.L. Rodin [l,k].

D. Riemann surfaces with zero boundary

Le M be some open Riemann surface and $\omega = \alpha dz(p) + \beta \overline{dz(p)}$ be a differential form. The Dirichlet integral of this form is defined as

$$D[\omega] = \iint_M \omega \wedge {*\omega} = \frac{1}{2i} \iint_M \{|\alpha|^2 + |\beta|^2\} \overline{dz(p)} \wedge dz(p),$$
$$*\omega = \frac{1}{2i} (\bar{\beta} dz(p) - \bar{\alpha} \overline{dz(p)}). \tag{11.49}$$

The Dirichlet integral of a function (or an integral) f is defined by the relation

$$D[f] = D[df]. \tag{11.50}$$

Let $L = \{l_n\}$ be some family of curves on M. The extremal length of this family is the value

$$\lambda(L) = \sup_{D[\omega]=1} \inf_{l_n \in L} \int_{l_n} |\omega| . \tag{11.51}$$

The surface M has a zero boundary (R. Nevanlinna [a]) if it possesses an exhaustion $M_n \to M$ ($M_n \subset M$ are domains of finite genus with a finite number of boundary contours, $\ldots \subset M_n \subset M_{n+1} \subset \ldots$) such that the extremal length of the family $\{\partial M_n\}$ tends to zero, $\lim_{n\to\infty} \lambda(\partial M_n) = 0$. The surfaces of this class which is denoted by O_G do not possess a Green function. The harmonic measure of the boundary of such a surface is equal to zero. The genus of $M \in O_G$ in general is equal to infinity.

We consider the Hilbert space H of first kind Abelian differentials with a finite Dirichlet integral. The space H possesses the basis $d\theta_1, d\theta_2, \ldots$ such that

$$\begin{aligned} &\operatorname{Im} \int_{K_{2j}} d\theta_{2l-1} = -\delta_{jl}, \quad \operatorname{Im} \int_{K_{2j-1}} d\theta_{2l-1} = 0, \\ &\operatorname{Im} \int_{K_{2j-1}} d\theta_{2l} = \delta_{jl}, \quad \operatorname{Im} \int_{K_{2j}} d\theta_{2l} = 0, \quad j,l = 1,2,\ldots . \end{aligned} \tag{11.52}$$

Moreover, below we use the Abelian differentials of the third kind $d\Omega_{q_0 q}(p)$ with residues ∓ 1 at the points $p = q_0, q$, respectively with periods

$$\operatorname{Re} \int_{K_j} d\Omega_{q_0 q_1}(p) = 0 \qquad j = 1,2,\ldots \tag{11.53}$$

having a finite Dirichlet integral over the domains $M' \subset M$ obtained by rejecting some compact neighbourhoods of the points q_0, q_1.

The following analogue of equation (5.13) is valid (Yu.L. Rodin [ ])

$$\begin{aligned} &\Omega_{s_0 s}(q) - \Omega_{s_0 s}(q_0) = \Omega_{q_0 q}(s) - \Omega_{q_0 q}(s_0) - \\ &- 2\pi i \lim_{n\to\infty} \sum_{j=1}^{g_n} \{\operatorname{Im} \int_{q_0}^{q} d\theta_{2j-1} \operatorname{Im} \int_{s_0}^{s} d\theta_{2j} - \operatorname{Im} \int_{q_0}^{q} d\theta_{2j} \operatorname{Im} \int_{s_0}^{s} d\theta_{2j-1}\} . \end{aligned} \tag{11.54}$$

Here g_n is the genus of M_n . By (11.51) there exists a subsequence $n_k \to \infty$ for which

$$\lim_{n_k\to\infty} \int_{l_{n_k}} |\omega| = 0 \ , \text{ for } D[\omega] = 1 \ . \tag{11.55}$$

The limits in (11.54) and below are taken along this subsequence. Below we will write $\sum\limits_1^\infty$

The Cauchy type integral kernel is determined, as in the compact case, by the relation

$$M(s,q) = \partial_s \, [\Omega_{s_0 s}(q) - \Omega_{s_0 s}(q_0)] \ . \tag{11.56}$$

By equation (11.54) we obtain the relation

$$M(s,q)dz(s) = d\Omega_{q_0 q}(s) - \\ - \pi \sum_1^\infty \{\mathrm{Im} \int_{q_0}^{q} d\theta_{2j-1} \cdot d\theta_{2j}(s) - \mathrm{Im} \int_{q_0}^{q} d\theta_{2j} \cdot d\theta_{2j-1}(s)\}. \tag{11.57}$$

The analyticity of the kernel with respect to q follows from (11.56) and the analyticity with respect to s follows from (11.57). From (11.56), (11.57) it follows that $M(s,q)$ is an Abelian covariant of the third kind with respect to s with poles in $s = q_0, q$ with residues ∓ 1 , respectively. The kernel is an Abelian integral of the second kind with poles of the first order at the point $q = s$ as a function of the variable q and $M(s,q_0) = 0$. Periods of the kernel along the cycles K_j $(j = 1,2,\ldots)$ of the homology basis are equal to (compare (5.20))

$$L_j(s) = \int_{K_j} d_q \, M(s,q) = \pi\theta_j'(s), \quad j = 1,2,\ldots \ . \tag{11.58}$$

From (11.57) it follows that the integral $M(s,q)$ for a fixed variable s has a finite Dirichlet integral over the surface M .

E. The Carleman-Bers-Vekua system

We suppose that $a,b \in L_r^{0,1}(M_0)$, $r > 2$, where $M_0 \subset M$ is compact and $a \equiv b \equiv 0$ outside M_0 . Consider the systems

$$\underset{\sim}{\partial} u \equiv \bar{\partial} u + au + b\bar{u} = 0 \ , \qquad (11.59)$$

$$\underset{\sim}{\partial}^* v \equiv - \bar{\partial} v + av + \overline{bv} = 0 \ . \qquad (11.60)$$

The operators

$$Pu = -\frac{1}{\pi} \iint_{M_0} [a(p)u(p) + b(p)\overline{u(p)}] M(p,q)\, d\sigma_p \ , \qquad (11.61)$$

$$P^*v = -\frac{1}{\pi} \iint_{M_0} [a(p)v(p) + \overline{b(p)}\,\overline{v(p)}] M(q,p)\, d\sigma_p \qquad (11.62)$$

are compact in the spaces $L_{q'}(M)$, and $L_{q'}^{1,0}(M)$, $\frac{1}{2} \le \frac{1}{r} + \frac{1}{q'}$, ≤ 1 , respectively. Therefore, the solvability conditions (6.11) are valid in the considered case and hence the construction of the Riemann-Roch theorem described above may be used in this case.

We formulate the final result.

Theorem 11.7. Let $M \in O_G$, $\gamma = \Sigma\, \alpha_k p_k \ge 0$, $\deg \gamma < \infty$, be some divisor and $H(\gamma)$ be the space of generalized analytic differentials of the first kind which are multiples of γ , K be the space of generalized analytic differentials with a single pole of first order at some point S_0 and $L(\gamma)$ be the space of generalized analytic functions which are multiples of the divisor $-\gamma$. Then

$$\dim(K/H(\gamma)) + \dim L(\gamma) = 2\deg\gamma + 2 \ . \qquad (11.63)$$

If the genus of M is finite, it is easy to show, that $\dim K = 2g$ and Theorem 11.7 is transmitted into the classical Riemann-Roch theorem.

§ 12. Some physical applications

A. Inverse scattering problem. Solitons

Consider the one-dimensional Schrödinger equation with real potential $u(x,t)$ (t is time)

$$L(t)\psi = -\frac{d^2\psi}{dx^2} + u(x,t)\psi = E\psi . \tag{12.1}$$

In order that the spectrum of the operator $L(t)$ is time independent the Lax equation

$$\dot{L} = [L,A] = LA - AL \tag{12.2}$$

has to be valid. Here A is an antisymmetric operator (B.M. Levitan [a]). Indeed, in this case

$$L(t) = \exp(-At)L(0)\exp(At) \tag{12.3}$$

and $L(t)$ is unitary equivalent to $L(0)$.

In the simplest case A is a differential operator of the third order,

$$A = 4\frac{d^3}{dx^3} - 3\left(u\frac{d}{dx} + \frac{d}{dx}u\right) \tag{12.4}$$

and (12.2) is the famous Korteweg de Vries equation (KdV)

$$u_t - 6uu_x + u_{xxx} = 0 \tag{12.5}$$

(see V.E. Zakharov, S.V. Manakov, S.P. Novikov, L.P. Pitaevsky [a], G.L. Lamb, jr. [a], B.M. Levitan [a]). If the order of the operator A is equal to $2n + 1$ $(n > 1)$, we have the so-called higher KdV equations.

If the potential u is known, one can construct the scattering theory for (12.1). Let $u(x,t)$ be a fast decreasing potential satisfying the condition

$$\int_{-\infty}^{\infty} (1+|x|)u(x,t)\,dx < \infty . \tag{12.6}$$

The Jost solutions of (12.1) are determined by the asymptotic behaviour

$$f_j(x,E) = \exp(-i\sqrt{E}x) + o(1), \quad x \to (-1)^j \infty, \quad j = 1,2 . \tag{12.7}$$

For real $E \geq 0$ the functions $f_j(x,E)$, $\overline{f_j(x,E)}$ form fundamental systems and hence

$$f_1(x,E) = a(E)f_2(x,E) + b(E)\overline{f_2(x,E)} . \tag{12.8}$$

The incident wave $\exp(-i\sqrt{E}x)$, the reflected wave $ba^{-1}\exp(i\sqrt{E}x)$ and the transmitted wave $a^{-1}\exp(-i\sqrt{E}x)$ are connected by the relations

$$a^{-1}(E)f_1(x,E) = \exp(-i\sqrt{E}x) + b(E)a^{-1}(E)\exp(i\sqrt{E}x) + o(1), \quad x \to +\infty,$$

$$a^{-1}(E)f_1(x,E) = a^{-1}(E)\exp(-i\sqrt{E}x) + o(1), \quad x \to -\infty . \tag{12.9}$$

The values $a^{-1}(E)b(E)$ and $a^{-1}(E)$ are called the reflection and transmission coefficients. The value $a(E)$ is called the scattering amplitude. From the relation

$$T(E) = \begin{bmatrix} a(E) & \overline{b(E)} \\ b(E) & \overline{a(E)} \end{bmatrix}, \det T(E) = \frac{W(f_1,\bar{f}_1)}{W(f_2,\bar{f}_2)}, \quad W(f_j,\bar{f}_j) = 2i\sqrt{E} \tag{12.10}$$

the formula

$$|a(E)|^2 - |b(E)|^2 = 1 \tag{12.11}$$

follows. Here $W(f,g)$ is the Wronskian determinant of the functions f and g. In the case considered $W(f,g)$ is independent of x.

It is known that the functions

$$g_j(x,E) = f_j(x,E)\exp(i\sqrt{E}x), \quad j = 1,2 \tag{12.12}$$

are analytic in the upper (for $j = 1$) and lower ($j = 2$) semi-planes of the plane $k = \sqrt{E}$, i.e. on the first $M^+(\mathrm{Im}\sqrt{E} > 0)$ and the second $M^-(\mathrm{Im}\sqrt{E} < 0)$ sheets of the Riemann surface M of the function $k^2 = E$. Since $W(f_1,\bar{f}_2) = a(E)$, the scattering amplitude is analytic on the plane M^+. It can be shown that $a(E)$ has a finite number of simple zeros at the points $E_j < 0$ $(j = 1,\dots,n)$ corresponding to the discrete spectrum of the operator L (bounded states).

Introduce the analytic vectors on $M^{\pm}$

$$\psi^{+}(E) = (g_1(x,E)\ ,\ \overline{g_2(x,\tilde{E})})\ ,\ \psi^{-}(E) = (g_2(x,E)\ ,\ \overline{g_1(x,\tilde{E})})\ . \tag{12.13}$$

Here $\sim$ is the operation of complex conjugation and projection on another sheet of M. We have on $L\{\mathrm{Re}\ E \geq 0\ ,\ \mathrm{Im}\ E = 0\}$

$$(f_1(x,E^{+})\ ,\ \overline{g_2(x,E^{-})}) = \frac{1}{a(E)}(f_2(x,E^{-}) + \overline{b(E)}f_1(x,E^{+})\ ,$$

$$-b(E)f_2(x,E^{-}) + \overline{f_1(x,E^{+})}) = (f_2(x,E^{-})\ ,\ \overline{f_1(x,E^{+})})G(E)\ ,$$

$$G(E) = \frac{1}{\overline{a(E)}}\left\|\begin{matrix} 1 & \overline{b(E)} \\ b(E) & 1 \end{matrix}\right\|\ . \tag{12.14}$$

Here $E^{\pm}$ are boundary points of $M^{\pm}$. If a point E^{+} belongs to the upper bank of the cut L^{+}, then the corresponding point E^{-} belongs to the lower bank of the cut L^{-} and conversely, $\tilde{E}^{\pm} = E^{\mp}$. By (12.12), (12.13) we have on L

$$\psi^{+}(E) = (f_1(x,E^{+})\ ,\ \overline{f_2(x,E^{-})})\exp\ (i\sqrt{E}x\sigma_3)\ ,$$

$$\psi^{-}(E) = (f_2(x,E^{-})\ ,\ \overline{f_1(x,E^{+})})\exp\ (i\sqrt{E}x\sigma_3)\ . \tag{12.15}$$

Here σ_3 is the Pauli matrix $\begin{bmatrix} 1 & 0 \\ 0 & -1 \end{bmatrix}$. We obtain the Riemann problem

$$\psi^{+}(E) = \psi^{-}(E)\exp\ (i\sqrt{E}x\sigma_3)G(E)\exp\ (-i\sqrt{E}x\sigma_3)\quad \text{on } L\ . \tag{12.16}$$

From (12.5) it follows that

$$\dot{a}(E) = 0\ ,\ \dot{b}(E) = 8i\ (\sqrt{E})^3 b(E)\ . \tag{12.17}$$

The discrete spectrum is determined by the eigenfunctions $f(x,E_j)$ $(j = 1,\ldots,n)$ with the asymptotic behaviour

$$f(x,E_j) = \exp\ (+\sqrt{-E_j}x) + o(1)\ ,\ x \to -\infty\ ,$$

$$f(x,E_j) = b_j(t)\exp\ (-\sqrt{-E_j}x) + o(1)\ ,\ x \to +\infty\ , \tag{12.18}$$

$$\dot{b}_j(t) = -\ 8i\ (\sqrt{-E_j})^3 b_j(t)\ ,\quad j = 1,\ldots,n\ .$$

The values $\{a(E), b(E), E_j, b_j, j = 1,\dots,n\}$ form a set called the scattering data.

The scattering data enable to re-establish the unknown potential $u(x,t)$ (the inverse scattering problem). For this purpose it is necessary to solve the problem (12.16) and then to calculate the potential by equation (12.1). For example, consider the Cauchy problem for equation (12.5). Taking into account the initial potential $u(x,0)$ we calculate the values $\{a(E), b(E), E_j, b_j, j = 1,\dots,n\}$ at $t = 0$ and then restore the scattering data dinamics by (12.17), (12.18). Now it is sufficient to solve the Riemann problem (12.16).

This problem is very simple in the reflectionless case $b(E) \equiv 0$. In this case instead of the problem (12.16) we have the scalar problem

$$a^{-1}(E)\psi^{+}(E) = \psi^{-}(E) . \tag{12.19}$$

The corresponding potential is called a soliton. We have no possibility to describe many interesting physical properties of solitons (see, for example, J.L. Lamb, jr. [a]).

If we refuse the condition (12.6) the situation becomes more complicated. For the periodic and almost periodic potentials there was constructed the theory of finite-zones integration (S.P. Novikov [a], S.P. Novikov, B.A. Dubrovin [a,b], B.A. Dubrovin [a,b], A.R. Its, V.B. Matveev [a,b], B.A. Dubrovin, V.B. Matveev, S.P. Novikov [a], V.A. Marchenko [a,b], P.D. Lax [a,b]) and later the case of infinite-zone spectrum was investigated (B.M. Levitan [a], H.P. McKean, E. Trubowitz [a,b]). We consider bounded differentiable real potentials $u(x,t)$ possessing the following property: the continuous spectrum of the operator $L(0)$ consists of a finite number of zones

$$L = \bigcup_{j=0}^{n-1} (E_{2j}, E_{2j+1}) \cup (E_{2n}, \infty) . \tag{12.20}$$

Let $\Theta(x,E)$, $\varphi(x,E)$ be the solution of equation (12.1) determined by the conditions

$$\begin{aligned} \Theta(0,E) &= \varphi'(0,E) = 1 , \\ \Theta'(0,E) &= \varphi(0,E) = 0 . \end{aligned} \tag{12.21}$$

In this case there exist functions $m_j(E)$, $j = 1,2$, called the Weyl's functions such that the solutions of equation (12.1)

$$f_j(x,E) = \Theta(x,E) + m_j(x,E)\varphi(x,E) \tag{12.22}$$

are square integrable over $(-\infty,0)$ (for $j = 1$) and $(0,\infty)$ $(j = 2)$ for any E, $\mathrm{Im}\, E \neq 0$. The Weyl's functions are represented in the form

$$m_j(E) = \int_{-\infty}^{\infty} \frac{d\rho_j(E')}{E-E'}, \quad j = 1,2, \tag{12.23}$$

where $\rho_j(E)$ is the spectral function of the operator $L(t)$ in the space $L_2(-\infty,0)$ (for $j = 1$) and $L_2(0,\infty)$ $(j = 2)$. From (12.23) one can see that poles of the functions $m_j(E)$ coincide with the discrete spectrum of the operator $L(t)$ in the spaces $L_2(-\infty,0)$ and $L_2(0,\infty)$. It can be shown that the functions $m_j(E)$ have a finite discontinuity along the continuous spectrum L (12.20) of the operator $L(t)$ in the space $L_2(-\infty,\infty)$ (B.M. Levitan, I.S. Sargsjan [a]). In the case (12.6) the Weyl-Titchmarsh functions (12.22) coincide up to factors depending on E with the Jost functions.

We have on L

$$\begin{gathered} f_1(x,E) = a(E)f_2(x,E) + b(E)\overline{f_2(x,E)}, \\ a(E) + b(E) = 1, \quad a(E)m_2(E) + b(E)\overline{m_2(E)} = m_1(E). \end{gathered} \tag{12.24}$$

The functions

$$g_j(x,E) = f_j(x,E)\exp\left(i\sqrt{E-E_{2n}}\,x\right) \tag{12.12'}$$

are analytic on the sheets $M^{\pm}$ of the surface of the function $z^2 = \prod_0^{2n}(E-E_j)$. Analogously to (12.15), (12.16) we obtain the Riemann problem

$$\psi^+(E) = \psi^-(E)\exp\left(i\sqrt{E-E_{2n}}\,x\sigma_3\right)G(E)\exp\left(-i\sqrt{E-E_{2n}}\,x\sigma_3\right) \quad \text{on } L \tag{12.16'}$$

on the Riemann surface M.

If $u(x,t)$ is periodic, it can be shown that $b \equiv 0$. In this case the function

$$B(E) = \begin{cases} f_1(x,E), & E \in M^+ \\ f_2(x,E), & E \in M^- \end{cases}, \tag{12.25}$$

coincides with the Baker-Akhiezer function.

The reader can find details in the expository paper Yu.L. Rodin [o]. The finite zones integration theory is expounded in the book V.E. Zakharov, S.V. Manakov, S.P. Novikov, L.P. Pitaevsky [a].

B. Integrable systems

Let $\psi(x,E)$ be a solution of equation (12.1). Introduce the function

$$\psi^*(x,E) = \psi_x(x,E) + i\sqrt{E}\, f(x,E) \tag{12.26}$$

and the vector $\Psi = \binom{\psi}{\psi^*}$. We obtain the system

$$\Psi_x = - i\sqrt{E}\Psi + \begin{pmatrix} 0 & 1 \\ u & 0 \end{pmatrix} \Psi . \tag{12.27}$$

Systems of the type

$$i\Psi_x = z\sigma_3\Psi + Q(x,t)\Psi \equiv L(x,t;z)\Psi \tag{12.28}$$

where Ψ is a 2×2 matrix and $Q(x,t)$ is off-diagonal is called a Zakharov-Shabat system. If

$$i\Psi_t = M(x,t;z)\Psi , \quad M = \sum_{j=0}^{n} z^j M_j(x,t) , \tag{12.29}$$

the system (12.28), (12.29) is compatible if

$$L_t - M_x + i[L,M] = 0 . \tag{12.30}$$

The Zakharov-Shabat equation (12.30) is a non-linear equation. Many equations playing an important role in the field theory, the solid state physics and other physical areas have the form (12.30). They are the KdV , nonlinear Schrödinger equations, Heisenberg and Landau-Lifschitz equations describing spin waves in ferromagnetic material, the famous sin-Gordon equation and so on. It is possible to reduce the inverse scattering problem for equation (12.28) to the Riemann problem as above, to restore its dynamics by (12.29) and to calculate the matrix potential $Q(x,t)$. This way can be used in that case when $L(x,t;z)$, $M(x,t;z)$ are analytic functions on an algebraic curve with respect to z (V.E. Zakharov, A.V. Mikhailov [a], see also I.M. Krichever, S.P. Novikov [a], A.V. Mikhailov [a,b], Yu.L. Rodin [m, n, o], E.K. Sklyanin [a]). Here we adduce another way proposed recently by R. Beals, R.R. Coifman [a,b] , A.S. Fokas, M.J. Ablowitz [a,b], M.J. Ablowitz, D. Bar Yaacov, A.S. Fokas [a] (see also V.E. Zakharov, S.V. Manakov [a]). We follow the paper of R. Beals, R.R. Coifman [b].

Let

$$\Phi(x,t;z) = \Psi(x,t;z)\exp(ixz\sigma_3) \ . \tag{12.31}$$

Then

$$D_z\Phi = -iQ\Phi \ , \ D_z = \frac{\partial}{\partial x} - iz[\cdot,\sigma_3] \ . \tag{12.32}$$

The operator D_z commutes with $\bar{\partial} = \frac{\partial}{\partial \bar{z}}$ and hence the matrix $\bar{\partial}\Phi$ is a solution of equation (12.32), too. Therefore

$$\bar{\partial}\Phi = \Phi A \tag{12.33}$$

where A is a solution of the homogeneous equation $D_z A = 0$, i.e.

$$A_x = iz(A\sigma_3 - \sigma_3 A) \ , \ A = \exp(-izx\sigma_3)A_0(z)\exp(izx\sigma_3) \tag{12.34}$$

where A_0 is an arbitrary matrix. It can be chosen so that the inverse scattering problem has the solution

$$Q(x,t) = [D_z,T](\Phi A) \tag{12.35}$$

where T is the operator (1.8). Here $A_0(z)$ is the "scattering data" .

APPENDIX

Cohomologies with coefficients in sheaves

1. Define for every open set U of points on a closed Riemann surface some module (group, vector space) $H(U)$ such that for each pair of open sets $U \subset V$ there is a homomorphism

$$\psi_U^V : H(V) \to H(U) \tag{A.1}$$

satisfying the condition

$$\psi_U^V \psi_V^W = \psi_U^W , \quad U \subset V \subset W .$$

This homomorphism is called a restriction of $H(V)$ on U. The family $P = \{H(U), \psi_U^V\}$ is called a projective system of modules (groups, vector spaces) or a presheaf over the surface M. Elements of $H(U)$ are called sections of the presheaf P over V, the group of sections is denoted by $\Gamma(P,U) = H(U)$

Let two presheaves $P = \{H(U), \psi_U^V\}$ and $P' = \{H'(U), \psi'^V_U\}$ over M be determined. A system of homomorphisms $r = \{r_U\}$, defined for every open set U

$$r_U : H(U) \to H'(U)$$

satisfying a commutativity condition $\psi'^V_U r_V = r_U \psi_U^V$, is called homomorphism of presheaves, $r: P \to P'$. In particular, if all r_U are monomorphisms, we obtain an embedding of presheaves. The factor-presheaf P/P' is defined by the vector-modules $H(U)/H'(U)$.

The homomorphisms sequence

$$\ldots \to P' \xrightarrow{\alpha} P \xrightarrow{\beta} P'' \to \ldots$$

is called exact if $\operatorname{Im} \alpha = \operatorname{Ker} \beta$. In particular, for any pair of presheaves $P' \subset P$ one can construct the exact sequence

$$0 \to P' \xrightarrow{i} P \xrightarrow{\pi} P'' \to 0 \tag{A.2}$$

where i is an embedding, π is a projection on the factor-presheaf $P'' = P/P'$. Such sequences are called short.

A presheaf P is called a sheaf if it satisfies two conditions.

a) Let $\{U_i\}$, $i \in I$, be some family of open sets on M, U be the union of these sets and $s_1, s_2 \in H(U)$. Then, if the restrictions of s_1 and s_2 on each set U_i belonging to $\{U_i\}$ coincide, it has to follow that $s_1 = s_2$.

b) Let $s_i \in H(U_i)$ be such that for any $i,j \in I$ the restrictions s_i and s_j are equal on $U_i \cap U_j$. Then there exists an $s \in H(U)$ the restriction on U_i of which is equal to s_i for any $i \in I$.

These notions are treated in many books (see, for example, R. Godement [a]). Take some examples of sheaves used in these books.

Differential forms $\omega = a dz$ and $\omega = a d\bar{z}$ are called of type $(1,0)$ and $(0,1)$, respectively. Differential forms $w = a dz \wedge d\bar{z}$ are called forms of type $(1,1)$. Forms of type $(0,0)$ (or 0) are functions on M.

Let $H(U)$ be a linear space of forms of type (i,j) of the class C^∞ in the domain U. The corresponding sheaf is called a sheaf of germs of differential forms of type (i,j) $(i,j = 0,1)$ and is denoted by $A^{i,j}$. We shall consider also subsheaves of these sheaves. $C^{i,j}$ is a sheaf of germs of closed differential forms of type (i,j). $\Omega^{i,j}$ is a sheaf of germs of holomorphic forms of type (i,j). Instead of $\Omega^{0,0}$ we will write also Ω.

Another example for a sheaf is given by the multiplicative group $M(U)$ of meromorphic functions in U. This sheaf is denoted by M^*. Its subsheaf formed by the germs of holomorphic functions different from zero is denoted by H^*.

2. Let $N = \{U_i, i \in I\}$ be some covering of the surface M by simply connected domains. Every finite set of indices $i_1,\dots,i_q$ determines the domain $U_{i_1,\dots,i_q} = U_{i_1} \cap \dots \cap U_{i_q}$. Consider some sheaf $P = \{H(U), \psi_U^V\}$ and associate to every domain $U_{i_0,\dots,i_q}$ some element $s_{i_0,\dots,i_q} \in H(U_{i_0,\dots,i_q})$. Such a correspondence defines a q-cochain $\{s_{i_0,\dots,i_q}\}$ with values in P. Finite linear combinations $\sum_k \alpha_k s_k^q$ of q-cochains s_k^q are q-cochains. Thus the set of q-cochains with values in P corresponding to a covering $\{U_i\}$ forms an abelian group $Z^q(P,\{U_i\})$. We define the coboundary operator

$$\delta\colon Z^q(P,\{U_i\}) \to Z^{q+1}(P,\{U_i\}), \quad \delta^2 = 0$$

by the formula

$$(\delta c)_{i_0,\dots,i_{q+1}} = \sum_{j=0}^{q+1} (-1)^j \psi^{U_{i_0,\dots,\hat{i}_j,\dots,i_{q+1}}}_{U_{i_0,\dots,i_q}} s_{i_0,\dots,\hat{i}_j,\dots,i_{q+1}} . \tag{A.3}$$

A cochain c with zero coboundary, $\delta c = 0$, is called a cocycle. The groups of q-cocycles are denoted by $Z^q(P,\{U_i\})$. A q-cocycle c that is the coboundary of some $(q-1)$-cochain $c', c = \delta c'$, is called cohomological to zero. Such cocycles form the group $B^q(P,\{U_i\}) = \delta Z^{q-1}(P,\{U_i\})$, $q > 0$, $B^0(P,\{U_i\}) \overset{\text{def.}}{=} 0$.

Groups of cohomologies of a covering N with coefficients in the sheaf P are defined by the relation

$$H^q(P,N) = Z^q(P,N)/B^q(P,N) . \tag{A.4}$$

Changing coverings it is possible to proceed to the projective limits called the cohomological group of the surface. But we will use only the group (A.4). The covering is fixed and the symbol $\{U_i\}$ will be omitted below. We shall write $H^q(P)$ or $H^q(P,M)$.

Consider a 0-cochain $\{s_i\}$ where the functions s_i are defined in the domains U_i. A coboundary of this cochain $\delta\{s_i\} = \{s_{ij}\}$ is the 1-cochain defined on intersections of domains

$$s_{ij} = s_i - s_j \quad \text{on} \quad U_i \cap U_j .$$

Therefore, if $\{s_i\}$ is a cocycle, $\delta(s_i) = 0$, then $s_i = s_j$ in all intersections $U_i \cap U_j$ and hence s_i forms one section over M. We obtain, hence, an important relation

$$H^0(P,M) = \Gamma(P,M) \tag{A.5}$$

where $\Gamma(P,M)$ is the group of sections of a sheaf over M. For the sheaf $A^{i,j}$ it is the group of smooth (i,j)-forms on the surface M.

Every short exact sequence of sheaves (A.2) is corresponding with the exact cohomological sequence

$$0 \to H^0(P') \xrightarrow{i^0} H^0(P) \xrightarrow{\pi^0} H^0(P'') \xrightarrow{\delta_0^*} H^1(P') \xrightarrow{i^1} H^1(P) \xrightarrow{\pi^1}$$

$$\to H^1(P'') \xrightarrow{\delta_1^*} \dots \to H^q(P') \xrightarrow{i^q} H^q(P) \xrightarrow{\pi^q} H^q(P'') \xrightarrow{\delta_q^*} \dots . \tag{A.6}$$

The induced homomorphisms i^q, π^q conform to homomorphisms of the corresponding groups of sections. They are defined on cohomological groups since they are commutative with the coboundary operator. Now we define the homomorphisms δ_q^*. Let a cocycle $\{s_{i_0,\dots,i_q}\}$ represent some element of the cohomological group $H^q(P'')$. If $P'' = (H''(U), \psi_U^V)$, then the value $s_{i_0,\dots,i_q} \in H''(U_{i_0,\dots,i_q})$ and is an image of some class of elements belonging to the group $H(U_{i_0,\dots,i_q})$ (see (A.2)). If $s'_{i_0,\dots,i_q}$, $s''_{i_0,\dots,i_q} \in H(U_{i_0,\dots,i_q})$ are two elements of this class being preimages of $s_{i_0,\dots,i_q}$, then $s'_{i_0,\dots,i_q} - s''_{i_0,\dots,i_q} = s^0_{i_0,\dots,i_q} \in H'(U_{i_0,\dots,i_q})$. Therefore, we have obtained the cochain $\{s^0_{i_0,\dots,i_q}\} \in Z^q(P')$. We define

$$\delta_q^* \{s_{i_0,\dots,i_q}\} = \delta \{s^0_{i_0,\dots,i_q}\} \in Z^{q+1}(P') .$$

Consider the short exact sequence

$$0 \to C \xrightarrow{i} A^0 \xrightarrow{d} C^1 \to 0 , \tag{A.7}$$

where C^1 is the sheaf of germs of closed differentials of the type $\omega = adz + bd\bar{z}$ and C is a constant sheaf. The exactness of the sequence (A.7) follows from the fact that in a simply connected domain every closed 1-form is exact and consequently the operator $d = \frac{\partial}{\partial z(p)} \wedge dz + \frac{\partial}{\partial \overline{z(p)}} \wedge d\bar{z}$ is an epimorphism on the space of smooth functions.

Because of (A.6) we have

$$\begin{aligned} 0 \to H^0(C,M) &\xrightarrow{i^0} H^0(A^0,M) \xrightarrow{d^0} H^0(C^1,M) \xrightarrow{\delta_0^*} \\ \to H^1(C,M) &\xrightarrow{i^1} H^1(A^0,M) \xrightarrow{d^1} H^1(C^1,M) \xrightarrow{\delta_1^*} \dots . \end{aligned} \tag{A.8}$$

Consider a partition of unity subordinated to the above fixed covering N, i.e. a set of infinitely differentiable functions $\alpha_i(p)$ the supports of which belong to the domains U_i such that for every $p \in M$ $\Sigma\, \alpha_i(p) \equiv 1$. The existence of such functions is proved in many books. The sheaf A^0 (and all sheaves $A^{i,j}$) possesses the following property. Let $\{f_i\}$ be a 0-cochain with values in A^0.

Then the 0-cochain $\{\alpha_i f_i\}$ is also a cochain with values in A^0. Sheaves possessing this property are called thin ones. Note that, for example, the sheaf C^1 is not thin. In fact, if $\{\omega_i\}$ is a 0-cochain with values in C^1, then forms $\omega_i\alpha_i$ cannot be closed.

<u>Theorem A.1</u>. For any thin sheaf P the relation

$$H^q(M,P) = 0 \; , \; q \geq 1 \tag{A.9}$$

is valid.

Firstly consider the case $q = 1$. Let f_{ij} be an arbitrary 1-cocycle with values in P. Consider the 0-cochain $f = \{f_i\}$,

$$f_i = \sum_k \alpha_k f_{ik}$$

where $\sum \alpha_j \equiv 1$ is a partition of unity. Since $\{f_{ij}\}$ is a cocycle, its boundary

$$\delta\{f_{ij}\} = \{f_{ij} - f_{ik} + f_{jk}\} = 0 \; .$$

Then a coboundary of f is

$$\delta f = \{f_i - f_j\} = \{\sum_k \alpha_k f_{ik} - \sum_k \alpha_k f_{jk}\} =$$

$$= \{\sum_k \alpha_k (f_{ik} - f_{jk})\} = \{\sum_k \alpha_k f_{ij}\} = \{f_{ij}\} \; .$$

Therefore $\{f_{ij}\} = \delta f$. Since f is a cochain with values in P, the cocycle $\{f_{ij}\}$ is cohomological to zero and hence $Z^1(P,M) = B^1(P,M)$ and $H^1(P,M) = 0$.

For $q > 1$ the same proof is valid too.

Taking into account (A.9) we obtain from (A.8) the exact sequences

$$\begin{gathered} 0 \to H^0(C,M) \xrightarrow{i^0} H^0(A^0,M) \xrightarrow{d^0} H^0(C^1,M) \xrightarrow{\delta^*} H^1(C,M) \to 0 \; , \\ 0 \to H^q(C^1,M) \xrightarrow{\delta^*_q} H^{q+1}(C,M) \to 0 \; . \end{gathered} \tag{A.10}$$

Therefore δ^*_q are epimorphisms for all $q > 0$. The following theorem is valid.

Theorem A.2 (de Rham).

$$H^1(C,M) \cong H^0(C^1,M)/d^0H^0(A^0,M) = \Gamma(C^1)/d\Gamma(A^0) ,$$
$$H^q(C^1,M) \cong H^{q+1}(C,M) . \tag{A.11}$$

The group $\Gamma(C^1)/d\Gamma(A^0)$ is known as the de Rham cohomology group. It is a factor group of closed differential forms by exact ones. Therefore, elements of this group are classes of closed forms having the same periods and hence this group is isomorphic to C^{2g} .

Let $h \in H^1(C)$. Construct a differential form representing the class corresponding to h in the de Rham group. To the cohomology class h there corresponds a 1-cocycle $\{h_{ij}\}$, where h_{ij} are some constants defined on intersections $U_i \cap U_j$. Because $\{h_{ij}\}$ is a cocycle, its coboundary is equal to zero

$$h_{ij} - h_{ik} + h_{jk} = 0 . \tag{A.12}$$

Let $\Sigma \alpha_k \equiv 1$ be a partition of unity subordinated to the covering $\{U_i\}$. Assume

$$H_i = \sum_{U_i \cap U_k \neq \emptyset} \alpha_k(p) h_{ik} . \tag{A.13}$$

$\{H_i\}$ is a 0-cochain with values in A^0 . On account of (A.12) we obtain

$$dH_i - dH_j = d(\sum_k \alpha_k)\, h_{ij} = 0 \tag{A.14}$$

since h_{ij} are constants. Therefore dH_i is a closed differential form corresponding to the cocycle h . Actually, let some differential form ω on M represent a cohomological class of the group $H^0(C^1)$. Accordingly to the construction presented in the proof of Theorem A.2 the coboundary homomorphism is built as follows. In the domain U_i we have $\omega = d\Omega_i$ where Ω_i is some function defined in U_i . The coboundary of $\{\Omega_i\}$ is the cocycle $\{\Omega_{ij}\}$, $\Omega_{ij} = \Omega_i - \Omega_j$, with values in C . It is clear that in this construction the form dH_i determines the cocycle $\{h_{ij}\}$.

In conclusion, introduce the value called Euler characteristic of the sheaf P

$$\chi(P) = \sum_{j=0}^{\infty} (-1)^j \dim H^j(M,P) \quad . \tag{A.15}$$

As follows from equation (A.15) for the short exact sequence

$$0 \to P' \to P \to P'' \to 0$$

there is the relation

$$\chi(P) = \chi(P') + \chi(P'') \quad . \tag{A.16}$$

Editor's Note:

This appendix may serve as a review of the basics of sheaf cohomology. It is appropiate for those readers who already know the theory. Those who do not, may benefit from reading a more detailed introduction to the subject, e.g. [ii], chapters 6,7,8, and 14, or [v], parts of ch. VII. This appendix also serves to familiarize the reader with the author's notation.

Notice the author's definition of cohomology (formula A.4 below): It uses a fixed covering of the manifold, rather than the usual procedure to be found in the above references (this involves further refinements of the covering plus taking a direct limit). The groups obtained using both procedures are, however, the same, i.e., they are isomorphic, if the fixed covering has been appropriately chosen. On this, see, e.g., p. 89 in L. Bers' reference [ii]. An example of such an appropriate covering is any locally finite one that, in addition, has this property: Whenever finitely many of the open sets of the covering have a non empty intersection, then such intersection is homeomorphic to an open ball of $\mathbb{R}^n$ (n = real dimension of the manifold). The existence of such a covering is guaranteed whenever the manifold is paracompact. See the references above. Any Riemann surface, i.e., any connected complex one-dimensional analytic manifold, is paracompact. An accessible and enlightening proof of this property of Riemann surfaces can be found in [i], p. 149.

NOTATIONS

$A^{i,j}(G)$ - the sheaf of germs of forms of the type (i,j) of the class class $L_p^{i,j}(G)$. (§ 3A).

$\hat{A}^{i,j}(G)$ - the sheaf of germs of forms of the type (i,j) of the class $\tilde{A}_p(G)$. (§ 3A).

$A^{i,j}(B,G)$ - the sheaf of germs of forms of the type (i,j) - sections of a bundle B of the class $L_p^{i,j}(G)$. (§ 3B).

$\hat{A}^{i,j}(B,G)$ - the sheaf of germs of forms of the type (i,j) - sections of a bundle B of the class $\tilde{A}_p(G)$. (§ 3B).

$\Omega(G)$ - the sheaf of germs of holomorphic functions in the domain G .

$\Omega^*(G)$ - the multiplicative sheaf of holomorphic functions different from zero in G .

$Q(\underset{\sim}{\partial},G)$ - the sheaf of germs of generalized analytic functions in the domain G . (§ 3A).

$Q(\underset{\sim}{\partial}_B,G)$ - the sheaf of germs of generalized analytic sections of the bundle B over G . (§ 3B).

$Q_\gamma(\underset{\sim}{\partial},G) = Q(-B_\gamma,G) = Q(\underset{\sim}{\partial}_{-B_\gamma},G)$ - the sheaf of germs of generalized analytic functions which is a multiple of the divisor - γ over G . (§ 3B).

$C^0(\bar{G})$, $C_\alpha^0(\bar{G})$, $L_{p,\alpha}(\bar{G})$, $C_{m,\alpha}(\bar{G})$, $C^{i,j}(\bar{G})$, $C_\alpha^{i,j}(\bar{G})$, $C_{m,\alpha}^{i,j}(\bar{G})$ - the spaces with norms (1.12), (1.13), (1.14), (2.3), (2.2_1) - (2.2_5) .

$D_{m,0}(\bar{G})$, $D_m(\bar{G})$, $D_{m,\infty}(\bar{G})$ - classes of functions possessing generalized derivatives of order $\le m$ (see (1.20)).

$A^*(G)$, $A(G)$ - the classes of meromorphic and holomorphic function in the domain G .

$\tilde{A}^*(a,b,F,G)$, $\tilde{A}(a,b,F,G)$, $A^*(a,b,F,G)$, $\tilde{A}_p(a,b,F,G)$, $\tilde{A}_p^*(a,b,G)$, $\tilde{A}_p(a,b,G)$, $A_p(a,b,G)$, $\tilde{A}_p(G)$, $A_p(G)$ - classes of functions related with the operator $\underset{\sim}{\partial}$ (pages 7 and 16).

$\tilde{A}^*{}_p^1$, $\tilde{A}_p^1$ - corresponding classes of covariants (page 16).

$L(\gamma)$, $H(\gamma)$ - the spaces of analytic (generalized analytic) functions and differentials which are multiples of the divisors - γ and γ , respectively. (§ 5A).

$S(d,\alpha,G)$, $S(\alpha)$ - spaces with the norm (11.11).

dw_j , $d\theta_j$, dt_p , dT_p , $d\omega_{q_0 q}$, $d\Omega_{q_0 q}$ - normalized Abelian differentials (5.1)-(5.13).

$K_j (j = 1,\ldots,2g)$ - the canonical homology basis of M (§ 5).

g - genus of the surface M .

$\underset{\sim}{\partial}$, $\underset{\sim}{\partial}^*$, $\underset{\sim}{\partial}_B$, $\underset{\sim}{\partial}_{B_\gamma}$ - differential operators (1.25), (2.7), (2.9), (3.14), (3.19).

$H^k(P)$ - the cohomology groups (§ 3, Appendix).

B_γ - the line bundle determined by the divisor γ (§ 3B) .

R E F E R E N C E S

1. Abdulaev R.N.
   a) On the solvability condition of the homogeneous Riemann problem on closed Riemann surfaces. Soviet Math. Doklady 4 (1963), 1525-1528.
   b) Zur Lösbarkeitsbedingung des homogenen Riemannschen Problems auf geschlossenen Riemannschen Flächen. Soobščenija Akad. Nauk Gruzin. SSR 35 (1964), 519-522 [Russisch].

2. Ablowitz M.J.; Bar Yaacov D.; Fokas A.S.
   a) On the inverse scattering transform for the Kadomtsev-Petviashvili equation. Stud. Appl. Math. 69 (1983), 135-143.

3. Ahlfors L.; Sario L.
   a) Riemann Surfaces. Princeton Mathematical Series, No. 26. Princeton Univ. Press., Princeton, N.J. 1960.

4. Beals R.W.; Coifman R.R.
   a) Scattering and inverse scattering for first order systems. Comm. Pure Appl. Math. 37 (1984), 39-90.
   b) Multidimensional scattering and inverse scattering. Yale Univ., preprint, 1984.

5. Behnke H.; Stein K.
   a) Entwicklung analytischer Funktionen auf Riemannschen Flächen. Math. Ann. 120 (1949) 430-461.

6. Bers L.
   a) Partial differential equations and generalized analytic functions. Proc. Nat. Acad. Sci. U.S.A. 37 (1951), 42-47.
   b) Theory of pseudo-analytic functions. Institute for Mathematics and Mechanics. New York University. New York 1953.
   c) Partial differential equations and pseudo-analytic functions on Riemann surfaces. Contributions to the theory of Riemann surfaces. Annals of Mathematics Studies. No. 30, 157-165. Princeton University Press, Princeton, N.J. 1953.

7. Bers L.; John F.; Schecher M.
   a) Partial differential equations. Interscience Publ., New York, 1964.

8. Bers L.; Nierenberg L.
   a) On a representation theorem for linear elliptic systems with discontinuous coefficients and its applications. Convegno Internazionale sulle Equazioni Lineari alle Derivate Parziali. Trieste 1954, 111-140. Edizioni Cremonese, Roma, 1955.
   b) On linear and non-linear elliptic boundary value problems in the plane. Convegno Internazionale sulle Equazioni Lineari alle Derivate Parziali. Trieste 1954, 141-167 Edizioni Cremonese Roma, 1955.

9. Bojarskiǐ B.V.
   a) Über ein Randwertproblem in der Funktionentheorie. Doklady Akad. Nauk SSSR 119 (1958), 199-202 [Russisch].

10. Carleman T.
a) Sur la théorie des équations intégrales et ses applications. Verh. internat. Math.-Kongr. 1 (1932), 138-151.

b) Sur les systémes linéaires aux dérivées partielles du premier ordre à deux variables. C.R. Acad. Sci. Paris 197 (1933), 471-474.

11. Calderon P.; Zygmund A.
a) On singular integrals. Amer. J. Math. 78 (1956), 289-309.

12. Cartan H.
a) Variétés analytiques complexes et cohomologie. Colloque sur les functions de plusieurs variables, tenu à Bruxelles, 1953, 41-55. Georges Thone, Liège, Masson & Cie., Paris, 1953.

13. Čibrikova L.I.
a) Das Riemannsche Randwertproblem für automorphe Funktionen im Fall von Gruppen mit zwei Invarianten. Izvestija vysš. učebn. Zaved. Mat. 1961, Nr. 6 (25) (1961), 121-131. Brief an die Redaktion Ibid. 1962 Nr. 3 (28) (1962), 195-196 [Russisch].

14. Dubrovin B.A.
a) The inverse scattering problem for periodic short-range potentials. Funkcional Anal. i Priložen, 9 (1975) No. 1, 65-66 [Russisch].
b) Endlichzonale lineare Differentialoperatoren und Abelsche Mannigfaltigkeiten. Uspehi Mat. Nauk 31 (190) (1976), 259-260 [Russisch].

15. Dubrovin B.A.; Matveev V.B.; Novikov S.P.
a) Nichtlineare Gleichungen Korteweg-de Vriesschen Typs, endlichzonige lineare Operatoren und Abelsche Mannigfaltigkeiten. Uspehi Mat. Nauk 31 (187) (1976), 55-136 [Russisch].

16. Fokas A.S.; Ablowitz M.J.
a) Comments on the inverse scattering transform and related nonlinear evolution equations. Nonlinear phenomena (Oaxtepec, 1982) 3-24. Lecture Notes in Phys. No. 189, Springer, Berlin-New York, 1983.
b) The inverse scattering transform for multidimensional (2+1) problems. Nonlinear phenomena (Oaxtepec, 1982) 137-183. Lecture Notes in Phys. No. 189, Springer, Berlin-New York, 1983.
c) On the inverse scattering of the time-dependent Schrödinger equation and the associated Kadomtsev-Petviashvili equation. Stud. Appl. Math. 69 (1983) No. 3, 211-228.

17. Fomenko V.T.; Tjurikov E.V.
a) On the bending of surfaces of genus $p > 0$ with boundary in a space of constant curvature under external constrains. Soviet Math. Doklady 17 (1976), 1527-1530 (1977).

18. Fomenko V.T.; Clymentov S.V.
a) Nonbendability of closed surfaces of genus $p \geq 1$ and positive extrinsic curvature. Math. USSR Sbornik 30 (1976), 361-372 (1978).

19. Forster O.
a) Riemannsche Flächen. Heidelberger Taschenbücher, Band 184, Springer-Verlag, Berlin-New York, 1977.

20. Gunning R.C.
   a) Lectures on Riemann surfaces. Mathematical Notes. Princeton University Press, Princeton, N.J. 1966.

21. Gunning R.C.; Rossi H.
   a) Analytic functions of several complex variables. Prentice-Hall. Inc., Englewood Cliffs, N.J. 1965.

22. Gahov F.D.
   a) Boundary-value problems. Pergamon Press, Oxford-New York-Paris; Addison-Wesley Publishing Co., Inc., Reading, Mass.-London, 1966.

23. Godement R.
   a) Topologie algébrique et théorie des faisceaux. Actualités Sci. Ind. No. 1252. Publ. Math. Univ. Strasbourg No. 13 Hermann, Paris 1958.

24. Gohberg I.C.; Kreĭn M.C.
   a) Systems of integral equations on the half-line with kernels depending on the difference of the arguments. Uspehi Math. Nauk (N.s.) 13 (1958) No. 2 (80), 3-72 [Russian].

25. Grothendieck A.
   a) Sur la classification des fibrés holomorphes sur la sphère de Riemann. Amer. J. Math. 79 (1957), 121-138.

26. Gusman S.Ja.; Rodin Yu.L.
   a) The kernel of an integral of Cauchy type on closed Riemann surfaces. Sibirsk. Mat. Ž. 3 (1962), 527-531 [Russian].

27. Hirzebruch F.
   a) Topological methods in algebraic geometry. Die Grundlehren der Mathematischen Wissenschaften. Band 131. Springer-Verlag, New York, 1966.

28. Hörmander L.
   a) An introduction to complex analysis in several variables. North-Holland, Amsterdam, 1973.

29. Its A.R.; Matveev V.B.
   a) Hill operators with a finite number of lacunae. Funktional Anal. i. Priložen. 9 (1975), 69-70 [Russian].
   b) Schrödinger operator with the finite-band spectrum and the N-soliton solutions of the Korteweg-de Vries equation. Mat. Fiz. 23 (1975), 51-68 (Russian, English summary).

30. Krasnoselskiĭ M.A.
   a) Topological methods in the theory of nonlinear integral equations. A Pergamon Press Book. The MacMillan Co., New York, 1964.

31. Krichever I.M., Novikov S.P.
   a) Holomorphic bundles and nonlinear equation. Physica D. 1981, Vol. 3, 267-293.

32. Koppelman W.
   a) Boundary value problems for pseudoanalytic functions. Bull. Amer. Math. Soc. 67 (1961), 371-376.

b) The Riemann-Hilbert problem for finite Riemann surfaces. Comm. Pure Appl. Math. 12 (1959), 13-35.
c) Singular integral equations, boundary value problems and the Riemann-Roch theorem. J. Math. Mech. 10 (1961), 247-277.

32. Lamb G.L. jr.
a) Elements of soliton theory. J. Wiley-Sons, New York, 1980.

33. Lax P.D.
a) Periodic solutions of the KdV equations. Nonlinear wave motion (Proc. AMS-SIAM Summer Sem., Clarkson Coll. Tech., Potsdam N.Y., (1972), 85-96. Lectures in Appl. Math., Vol. 15, Amer. Math. Soc., Providence, R.I., 1974.
b) Periodic solutions of the KdV equation. Comm. Pure Appl. Math. 28 (1975), 141-188.

34. Levitan B.M.
a) Inverse Sturm-Liouville problems. Moscow, Nauka, 1984 [Russian].

35. Levitan B.M.; Sargasjan I.S.
a) Introduction to spectral theory: selfadjoint ordinary differential operators. Translations of Mathematical Monographs. Vol. 39. American Mathematical Society, Providence, R.I., 1975.

36. Marchenko V.A.
a) A periodic Korteweg-de Vries problem. Dokl. Akad. Nauk SSSR 217 (1974), 276-279 [Russian].
b) The periodic Korteweg-de Vries problem. Matem. Sbornik. (N.S.) 95 (137) (1974), 331-356 [Russian].

37. McKean H.P.; Trubowitz E.
a) Hill's operator and hyperelliptic function theory in the presence of infinitely many branch points. Comm. Pure Appl. Math. 29 (1976), 143-226.
b) Hill's surfaces and their theta functions. Bull. Amer. Math. Soc. 84 (1978), 1042-1085.

38. Mikhailov A.V.
a) The reduction problem and the inverse scattering method. Physica D 1981, Vol. 3, 73-117.
b) The Landau-Lifschitz equation and the Riemann boundary problem on a torus. Phys. Lett. A 92 (1982), 51-55.

39. Mikhailov L.G.
a) New class of singular integral equations and its applications to differential equations with singular coefficients. Dushanbe 1963 [Russian].

40. Muskhelishvili N.I.
a) Singular integral equations. Noordhoff, Groningen, 1953.

41. Nevanlinna R.
a) Uniformisierung. Die Grundlehren der Mathematischen Wissenschaften in Einzeldarstellungen mit besonderer Berücksichtigung der Anwendungsgebiete. Band LXIV. Springer-Verlag, Berlin-Göttingen-Heidelberg, 1953.

42. Novikov S.P.
a) A periodic problem for the Korteweg-de Vries equation I. Funkcional Anal. i Priložen 8 (1974), 54-66 [Russian].

43. Novikov S.P.; Dubrovin B.A.
a) A periodic problem for the Korteweg-de Vries and Sturm-Liouville equations. Their connection with algebraic geometry. Dokl. Akad. Nauk SSSR 219 (1974), 531-534 [Russian].
b) Periodic and conditionally periodic analogs of the many-soliton solutions of the Korteweg-de Vries equation. Ž. Eksper Teoret. Fiz. 67 (1974), No. 6, 2131-2144. Engl. Translation. Soviet physics JETP 40 (1974) No. 6, 1058-1063.

44. Prößdorf S.
a) Einige Klassen singulärer Gleichungen. Mathematische Reihe, Band 46, Birkhäuser Verlag, Basel-Stuttgart, 1974.

45. Rodin Y.L.
a) Conditions for the solvability of Riemann's and Hilbert's boundary value problems on Riemannian surfaces. Dokl. Akad. Nauk SSSR 129 (1959), 1234-1237 [Russian].
b) The characteristic functions of certain integral equations. Dokl. Akad. Nauk SSSR 130 (1960), 23-25 [Russian]; translated in Soviet Math. Dokl. 1 (1960), 13-15.
c) On the Riemann problem on closed Riemann surfaces. Soviet Math. Dokl. 1 (1960), 723-725.
d) Certain problems in Russian Mathematics and Mechanics. In honour of M.A. Lavrent'ev: SB USSR Acad. sei Novosibirsk, 1961, 224-226.
e) Algebraic theory of generalized analytic functions on closed Riemann surfaces. Dokl. Akad. Nauk SSSR 142 (1962), 1030-1033. [Russian]; translated in Soviet Math. Dokl. 3 (1962), 243-246.
f) Integrals of Cauchy type and boundary value problems for generalized analytic functions on closed Riemann surfaces. Dokl. Akad Nauk SSSR 142 (1962), 798-801 [Russian]; translated Soviet Math. Dokl. 3 (1962), 177-181.
g) On the Riemann boundary-value problem with discontinuous coefficients on Riemann surfaces. Perm. Gos. Univ. Učep. Zap. Mat. 17 (1960), No. 2, 79-81 [Russian].
The Riemann boundary-value problem for differentials on closed Riemann surfaces. Perm. Gos. Univ. Učep. Zap. Mat. 17 (1960) No. 2, 83-85 [Russian].
h) On the algebraic theory of elliptic systems of first-order differential equations. Soviet Math. Doklady 4 (1963), 868-871.
i) The elliptic operators of first order on Riemann surfaces. Intern. Math. Congress Abstr. Sec. 10, 16, Moscow, 1966.
j) On the theory of many-valued generalized analytic functions. Sakharth SSR Mecn. Akad. Moambe 43 (1966), 261-268 (Russian, Georgian Summary).
k) The second Cousin problem on Riemann surfaces of infinite genus. Soviet Math. Dokl. 13 (1972), 550-554.
l) Nonlinear problems of the theory of functions on open Riemann surfaces. Some problems in modern function theory (Proc. Conf. Modern Problems of Geometric Theory of Functions. Inst. Math. Acad. Sci. USSR Novosibirsk) 1976, 111-118 [Russian]. Akad. Nauk SSSR Sibirsk. Otdel, Inst. Mat. Novosibirsk, 1976.
m) The Riemann boundary problem on a torus and the inverse scattering problem for the Landau-Lifschitz equation. Lett. Math. Phys. 7 (1983), 3-8.

n) The Riemann boundary problem on Riemann surfaces and the inverse scattering problem for the Landau-Lifschitz equation. Physica 11D (1984), 90-108.
o) The Riemann boundary value problem on closed Riemann surfaces and integrable systems. Physica 24D (1987), 1-53.
p) The structure of the general solution of the Riemann boundary value problem for a holomorphic vector on a compact Riemann surface. Soviet Math. Dokl. 18 (1977), 201-205.

46. Rodin Ju.L.; Turakulov A.
a) The Riemann boundary value problem for generalized analytic functions with singular coefficients on a compact Riemann surface. Soobšč. Acad. Nauk Gruzin. SSR 96 (1976), 21-24 (Russian; Georgian and English summaries).

47. Röhrl H.
a) Über das Riemann-Privalovsche Randwertproblem. Math. Ann. 151 (1963), 365-423.
b) Ω-degenerate singular integral equations and holomorphic affine bundles over compact Riemann surfaces. I. Comment. Math. Helv. 38 (1963), 84-120.

48. Serre J.P.
a) Quelques problèms globaux relatifs aux variétés de Stein. Colloque sur les fonctions de plusieurs variables, tenu à Bruxelles, 1953, 57-68. Georges Thone, Liège, Masson & Cie, Paris, 1953.

49. Springer G.
a) Introduction to Riemann surfaces. Addison-Wesley Publishing Company Inc., Reading, Mass., 1957.

50. Sklyanin E.K.
a) On complete integrability of the Landau-Lifschitz equation. Preprint LOMI E-3-79. Leningrad, 1979.

51. Théodoresco N.
a) La dérivée aréolaire et ses applications à la physique mathématique. Paris, Diss., 1931.
b) La dérivée aréolaire. Ann. Roum. Math. Cahier 3, (1936), 3-62.

52. Tietz H.
a) Fabersche Entwicklungen auf geschlossenen Riemannschen Flächen. J. Reine Angew. Math. 190 (1952), 22-33.

53. Turakulov A.
a) Der Riemann-Rochsche Satz für verallgemeinerte analytische Funktionen mit singulären Koeffizienten. Dokl. Akad. Nauk UZSSR, 1975, No. 8, (1975), 11-12 [Russisch].
b) Generalized constants for the Carleman system with singular coefficients. Problems of Mathematics. Tashkent State Univ. 1975, 461, 25.

54. Vekua I.N.
a) Generalized analytic functions. Pergamon Press, London-Paris-Frankfurt; Addison-Wesley Publishing Co., Inc, Reading, Mass.; 1962.

b) Systems of differential equations of the first order of elliptic type and boundary value problems with an application to the theory of shells. Mat. Sbornik N.S. 31 (73) (1952), 217-314 [Russian].

55. Volkoviskiǐ K.L.
a) Generalized analytic functions on open Riemann surfaces. Soviet Math. Dokl. Vol. 16 (1975), 1443-1446.

56. Wendland W.L.
a) Elliptic systems in the plane. Pitman, London-San Francisco-Melbourne, 1979.

57. Zakharov V.Z.; Manakov S.V.
a) Construction of multi-dimensional non-linear integrable systems and their solutions. Functional Anal. i Priložen 19 (1985), 11-25.

58. Zakharov V.E.; Manakov S.V.; Novikov S.P.; Pitaevsky L.P.
a) Theory of solitons. The method of the inverse problem. Nauka, Moscow, 1980 [Russian].

59. Zakharov V.E.; Mikhailov A.V.
a) The method of the inverse scattering problem with spectral parameter on an algebraic curve. Funkcional Anal. i Priložen 17 (1983), 1-6 [Russian].

REFERENCES TO APPENDIX

i. Ahlfors, L.: Conformal invariants. McGraw-Hill, New York etc., 1973.

ii. Bers, L.: Introduction to several complex variables. Courant Inst. of Math. Sci., Notes. New York, 1964.

iii. Forster, O.: Riemannsche Flächen. Springer-Verlag, Berlin-Heidelberg-New York, 1977.

iv. Godement, R.: Topologie algébrique et théorie des faisceaux, Hermann, Paris, 1958.

v. Hörmander, L.: An introduction to complex analysis in several variables, 2nd ed. . North Holland, Amsterdam-London, 1979.

vi. Husemoller, D.: Fibre bundles. McGraw-Hill, New York etc., 1966.

INDEX

Vol. 1117: D.J. Aldous, J.A. Ibragimov, J. Jacod, Ecole d'Été de Probabilités de Saint-Flour XIII – 1983. Édité par P.L. Hennequin. IX, 409 pages. 1985.

Vol. 1118: Grossissements de filtrations: exemples et applications. Seminaire, 1982/83. Edité par Th. Jeulin et M. Yor. V, 315 pages. 1985.

Vol. 1119: Recent Mathematical Methods in Dynamic Programming. Proceedings, 1984. Edited by I. Capuzzo Dolcetta, W.H. Fleming and T. Zolezzi. VI, 202 pages. 1985.

Vol. 1120: K. Jarosz, Perturbations of Banach Algebras. V, 118 pages. 1985.

Vol. 1121: Singularities and Constructive Methods for Their Treatment. Proceedings, 1983. Edited by P. Grisvard, W. Wendland and J.R. Whiteman. IX, 346 pages. 1985.

Vol. 1122: Number Theory. Proceedings, 1984. Edited by K. Alladi. VII, 217 pages. 1985.

Vol. 1123: Séminaire de Probabilités XIX 1983/84. Proceedings. Edité par J. Azéma et M. Yor. IV, 504 pages. 1985.

Vol. 1124: Algebraic Geometry, Sitges (Barcelona) 1983. Proceedings. Edited by E. Casas-Alvero, G.E. Welters and S. Xambó-Descamps. XI, 416 pages. 1985.

Vol. 1125: Dynamical Systems and Bifurcations. Proceedings, 1984. Edited by B.L.J. Braaksma, H.W. Broer and F. Takens. V, 129 pages. 1985.

Vol. 1126: Algebraic and Geometric Topology. Proceedings, 1983. Edited by A. Ranicki, N. Levitt and F. Quinn. V, 423 pages. 1985.

Vol. 1127: Numerical Methods in Fluid Dynamics. Seminar. Edited by F. Brezzi, VII, 333 pages. 1985.

Vol. 1128: J. Elschner, Singular Ordinary Differential Operators and Pseudodifferential Equations. 200 pages. 1985.

Vol. 1129: Numerical Analysis, Lancaster 1984. Proceedings. Edited by P.R. Turner. XIV, 179 pages. 1985.

Vol. 1130: Methods in Mathematical Logic. Proceedings, 1983. Edited by C.A. Di Prisco. VII, 407 pages. 1985.

Vol. 1131: K. Sundaresan, S. Swaminathan, Geometry and Nonlinear Analysis in Banach Spaces. III, 116 pages. 1985.

Vol. 1132: Operator Algebras and their Connections with Topology and Ergodic Theory. Proceedings, 1983. Edited by H. Araki, C.C. Moore, Ş. Strătilă and C. Voiculescu. VI, 594 pages. 1985.

Vol. 1133: K.C. Kiwiel, Methods of Descent for Nondifferentiable Optimization. VI, 362 pages. 1985.

Vol. 1134: G.P. Galdi, S. Rionero, Weighted Energy Methods in Fluid Dynamics and Elasticity. VII, 126 pages. 1985.

Vol. 1135: Number Theory, New York 1983–84. Seminar. Edited by D.V. Chudnovsky, G.V. Chudnovsky, H. Cohn and M.B. Nathanson. V, 283 pages. 1985.

Vol. 1136: Quantum Probability and Applications II. Proceedings, 1984. Edited by L. Accardi and W. von Waldenfels. VI, 534 pages. 1985.

Vol. 1137: Xiao G., Surfaces fibrées en courbes de genre deux. IX, 103 pages. 1985.

Vol. 1138: A. Ocneanu, Actions of Discrete Amenable Groups on von Neumann Algebras. V, 115 pages. 1985.

Vol. 1139: Differential Geometric Methods in Mathematical Physics. Proceedings, 1983. Edited by H. D. Doebner and J. D. Hennig. VI, 337 pages. 1985.

Vol. 1140: S. Donkin, Rational Representations of Algebraic Groups. VII, 254 pages. 1985.

Vol. 1141: Recursion Theory Week. Proceedings, 1984. Edited by H.-D. Ebbinghaus, G.H. Müller and G.E. Sacks. IX, 418 pages. 1985.

Vol. 1142: Orders and their Applications. Proceedings, 1984. Edited by I. Reiner and K. W. Roggenkamp. X, 306 pages. 1985.

Vol. 1143: A. Krieg, Modular Forms on Half-Spaces of Quaternions. XIII, 203 pages. 1985.

Vol. 1144: Knot Theory and Manifolds. Proceedings, 1983. Edited by D. Rolfsen. V, 163 pages. 1985.

Vol. 1145: G. Winkler, Choquet Order and Simplices. VI, 143 pages. 1985.

Vol. 1146: Séminaire d'Algèbre Paul Dubreil et Marie-Paule Malliavin. Proceedings, 1983–1984. Edité par M.-P. Malliavin. IV, 420 pages. 1985.

Vol. 1147: M. Wschebor, Surfaces Aléatoires. VII, 111 pages. 1985.

Vol. 1148: Mark A. Kon, Probability Distributions in Quantum Statistical Mechanics. V, 121 pages. 1985.

Vol. 1149: Universal Algebra and Lattice Theory. Proceedings, 1984. Edited by S. D. Comer. VI, 282 pages. 1985.

Vol. 1150: B. Kawohl, Rearrangements and Convexity of Level Sets in PDE. V, 136 pages. 1985.

Vol 1151: Ordinary and Partial Differential Equations. Proceedings, 1984. Edited by B.D. Sleeman and R.J. Jarvis. XIV, 357 pages. 1985.

Vol. 1152: H. Widom, Asymptotic Expansions for Pseudodifferential Operators on Bounded Domains. V, 150 pages. 1985.

Vol. 1153: Probability in Banach Spaces V. Proceedings, 1984. Edited by A. Beck, R. Dudley, M. Hahn, J. Kuelbs and M. Marcus. VI, 457 pages. 1985.

Vol. 1154: D.S. Naidu, A.K. Rao, Singular Pertubation Analysis of Discrete Control Systems. IX, 195 pages. 1985.

Vol. 1155: Stability Problems for Stochastic Models. Proceedings, 1984. Edited by V.V. Kalashnikov and V.M. Zolotarev. VI, 447 pages. 1985.

Vol. 1156: Global Differential Geometry and Global Analysis 1984. Proceedings, 1984. Edited by D. Ferus, R.B. Gardner, S. Helgason and U. Simon. V, 339 pages. 1985.

Vol. 1157: H. Levine, Classifying Immersions into $\mathbb{R}^4$ over Stable Maps of 3-Manifolds into $\mathbb{R}^2$. V, 163 pages. 1985.

Vol. 1158: Stochastic Processes – Mathematics and Physics. Proceedings, 1984. Edited by S. Albeverio, Ph. Blanchard and L. Streit. VI, 230 pages. 1986.

Vol. 1159: Schrödinger Operators, Como 1984. Seminar. Edited by S. Graffi. VIII, 272 pages. 1986.

Vol. 1160: J.-C. van der Meer, The Hamiltonian Hopf Bifurcation. VI, 115 pages. 1985.

Vol. 1161: Harmonic Mappings and Minimal Immersions, Montecatini 1984. Seminar. Edited by E. Giusti. VII, 285 pages. 1985.

Vol. 1162: S.J.L. van Eijndhoven, J. de Graaf, Trajectory Spaces, Generalized Functions and Unbounded Operators. IV, 272 pages. 1985.

Vol. 1163: Iteration Theory and its Functional Equations. Proceedings, 1984. Edited by R. Liedl, L. Reich and Gy. Targonski. VIII, 231 pages. 1985.

Vol. 1164: M. Meschiari, J.H. Rawnsley, S. Salamon, Geometry Seminar "Luigi Bianchi" II – 1984. Edited by E. Vesentini. VI, 224 pages. 1985.

Vol. 1165: Seminar on Deformations. Proceedings, 1982/84. Edited by J. Ławrynowicz. IX, 331 pages. 1985.

Vol. 1166: Banach Spaces. Proceedings, 1984. Edited by N. Kalton and E. Saab. VI, 199 pages. 1985.

Vol. 1167: Geometry and Topology. Proceedings, 1983–84. Edited by J. Alexander and J. Harer. VI, 292 pages. 1985.

Vol. 1168: S.S. Agaian, Hadamard Matrices and their Applications. III, 227 pages. 1985.

Vol. 1169: W.A. Light, E.W. Cheney, Approximation Theory in Tensor Product Spaces. VII, 157 pages. 1985.

Vol. 1170: B.S. Thomson, Real Functions. VII, 229 pages. 1985.

Vol. 1171: Polynômes Orthogonaux et Applications. Proceedings, 1984. Edité par C. Brezinski, A. Draux, A.P. Magnus, P. Maroni et A. Ronveaux. XXXVII, 584 pages. 1985.

Vol. 1172: Algebraic Topology, Göttingen 1984. Proceedings. Edited by L. Smith. VI, 209 pages. 1985.

Vol. 1173: H. Delfs, M. Knebusch, Locally Semialgebraic Spaces. XVI, 329 pages. 1985.

Vol. 1174: Categories in Continuum Physics, Buffalo 1982. Seminar. Edited by F.W. Lawvere and S.H. Schanuel. V, 126 pages. 1986.

Vol. 1175: K. Mathiak, Valuations of Skew Fields and Projective Hjelmslev Spaces. VII, 116 pages. 1986.

Vol. 1176: R.R. Bruner, J.P. May, J.E. McClure, M. Steinberger, H_∞ Ring Spectra and their Applications. VII, 388 pages. 1986.

Vol. 1177: Representation Theory I. Finite Dimensional Algebras. Proceedings, 1984. Edited by V. Dlab, P. Gabriel and G. Michler. XV, 340 pages. 1986.

Vol. 1178: Representation Theory II. Groups and Orders. Proceedings, 1984. Edited by V. Dlab, P. Gabriel and G. Michler. XV, 370 pages. 1986.

Vol. 1179: Shi J.-Y. The Kazhdan-Lusztig Cells in Certain Affine Weyl Groups. X, 307 pages. 1986.

Vol. 1180: R. Carmona, H. Kesten, J.B. Walsh, École d'Été de Probabilités de Saint-Flour XIV – 1984. Édité par P.L. Hennequin. X, 438 pages. 1986.

Vol. 1181: Buildings and the Geometry of Diagrams, Como 1984. Seminar. Edited by L. Rosati. VII, 277 pages. 1986.

Vol. 1182: S. Shelah, Around Classification Theory of Models. VII, 279 pages. 1986.

Vol. 1183: Algebra, Algebraic Topology and their Interactions. Proceedings, 1983. Edited by J.-E. Roos. XI, 396 pages. 1986.

Vol. 1184: W. Arendt, A. Grabosch, G. Greiner, U. Groh, H.P. Lotz, U. Moustakas, R. Nagel, F. Neubrander, U. Schlotterbeck, One-parameter Semigroups of Positive Operators. Edited by R. Nagel. X, 460 pages. 1986.

Vol. 1185: Group Theory, Beijing 1984. Proceedings. Edited by Tuan H.F. V, 403 pages. 1986.

Vol. 1186: Lyapunov Exponents. Proceedings, 1984. Edited by L. Arnold and V. Wihstutz. VI, 374 pages. 1986.

Vol. 1187: Y. Diers, Categories of Boolean Sheaves of Simple Algebras. VI, 168 pages. 1986.

Vol. 1188: Fonctions de Plusieurs Variables Complexes V. Séminaire, 1979–85. Edité par François Norguet. VI, 306 pages. 1986.

Vol. 1189: J. Lukeš, J. Malý, L. Zajíček, Fine Topology Methods in Real Analysis and Potential Theory. X, 472 pages. 1986.

Vol. 1190: Optimization and Related Fields. Proceedings, 1984. Edited by R. Conti, E. De Giorgi and F. Giannessi. VIII, 419 pages. 1986.

Vol. 1191: A.R. Its, V.Yu. Novokshenov, The Isomonodromic Deformation Method in the Theory of Painlevé Equations. IV, 313 pages. 1986.

Vol. 1192: Equadiff 6. Proceedings, 1985. Edited by J. Vosmansky and M. Zlámal. XXIII, 404 pages. 1986.

Vol. 1193: Geometrical and Statistical Aspects of Probability in Banach Spaces. Proceedings, 1985. Edited by X. Femique, B. Heinkel, M.B. Marcus and P.A. Meyer. IV, 128 pages. 1986.

Vol. 1194: Complex Analysis and Algebraic Geometry. Proceedings, 1985. Edited by H. Grauert. VI, 235 pages. 1986.

Vol. 1195: J.M. Barbosa, A.G. Colares, Minimal Surfaces in $\mathbb{R}^3$. X, 124 pages. 1986.

Vol. 1196: E. Casas-Alvero, S. Xambó-Descamps, The Enumerative Theory of Conics after Halphen. IX, 130 pages. 1986.

Vol. 1197: Ring Theory. Proceedings, 1985. Edited by F.M.J. van Oystaeyen. V, 231 pages. 1986.

Vol. 1198: Séminaire d'Analyse, P. Lelong – P. Dolbeault – H. Skoda. Seminar 1983/84. X, 260 pages. 1986.

Vol. 1199: Analytic Theory of Continued Fractions II. Proceedings, 1985. Edited by W.J. Thron. VI, 299 pages. 1986.

Vol. 1200: V.D. Milman, G. Schechtman, Asymptotic Theory of Finite Dimensional Normed Spaces. With an Appendix by M. Gromov. VIII, 156 pages. 1986.

Vol. 1201: Curvature and Topology of Riemannian Manifolds. Proceedings, 1985. Edited by K. Shiohama, T. Sakai and T. Sunada. VII, 336 pages. 1986.

Vol. 1202: A. Dür, Möbius Functions, Incidence Algebras and Power Series Representations. XI, 134 pages. 1986.

Vol. 1203: Stochastic Processes and Their Applications. Proceedings, 1985. Edited by K. Itô and T. Hida. VI, 222 pages. 1986.

Vol. 1204: Séminaire de Probabilités XX, 1984/85. Proceedings. Edité par J. Azéma et M. Yor. V, 639 pages. 1986.

Vol. 1205: B.Z. Moroz, Analytic Arithmetic in Algebraic Number Fields. VII, 177 pages. 1986.

Vol. 1206: Probability and Analysis, Varenna (Como) 1985. Seminar. Edited by G. Letta and M. Pratelli. VIII, 280 pages. 1986.

Vol. 1207: P.H. Bérard, Spectral Geometry: Direct and Inverse Problems. With an Appendix by G. Besson. XIII, 272 pages. 1986.

Vol. 1208: S. Kaijser, J.W. Pelletier, Interpolation Functors and Duality. IV, 167 pages. 1986.

Vol. 1209: Differential Geometry, Peñíscola 1985. Proceedings. Edited by A.M. Naveira, A. Ferrández and F. Mascaró. VIII, 306 pages. 1986.

Vol. 1210: Probability Measures on Groups VIII. Proceedings, 1985. Edited by H. Heyer. X, 386 pages. 1986.

Vol. 1211: M.B. Sevryuk, Reversible Systems. V, 319 pages. 1986.

Vol. 1212: Stochastic Spatial Processes. Proceedings, 1984. Edited by P. Tautu. VIII, 311 pages. 1986.

Vol. 1213: L.G. Lewis, Jr., J.P. May, M. Steinberger, Equivariant Stable Homotopy Theory. IX, 538 pages. 1986.

Vol. 1214: Global Analysis – Studies and Applications II. Edited by Yu.G. Borisovich and Yu.E. Gliklikh. V, 275 pages. 1986.

Vol. 1215: Lectures in Probability and Statistics. Edited by G. del Pino and R. Rebolledo. V, 491 pages. 1986.

Vol. 1216: J. Kogan, Bifurcation of Extremals in Optimal Control. VIII, 106 pages. 1986.

Vol. 1217: Transformation Groups. Proceedings, 1985. Edited by S. Jackowski and K. Pawalowski. X, 396 pages. 1986.

Vol. 1218: Schrödinger Operators, Aarhus 1985. Seminar. Edited by E. Balslev. V, 222 pages. 1986.

Vol. 1219: R. Weissauer, Stabile Modulformen und Eisensteinreihen. III, 147 Seiten. 1986.

Vol. 1220: Séminaire d'Algèbre Paul Dubreil et Marie-Paule Malliavin. Proceedings, 1985. Edité par M.-P. Malliavin. IV, 200 pages. 1986.

Vol. 1221: Probability and Banach Spaces. Proceedings, 1985. Edited by J. Bastero and M. San Miguel. XI, 222 pages. 1986.

Vol. 1222: A. Katok, J.-M. Strelcyn, with the collaboration of F. Ledrappier and F. Przytycki, Invariant Manifolds, Entropy and Billiards; Smooth Maps with Singularities. VIII, 283 pages. 1986.

Vol. 1223: Differential Equations in Banach Spaces. Proceedings, 1985. Edited by A. Favini and E. Obrecht. VIII, 299 pages. 1986.

Vol. 1224: Nonlinear Diffusion Problems, Montecatini Terme 1985. Seminar. Edited by A. Fasano and M. Primicerio. VIII, 188 pages. 1986.

Vol. 1225: Inverse Problems, Montecatini Terme 1986. Seminar. Edited by G. Talenti. VIII, 204 pages. 1986.

Vol. 1226: A. Buium, Differential Function Fields and Moduli of Algebraic Varieties. IX, 146 pages. 1986.

Vol. 1227: H. Helson, The Spectral Theorem. VI, 104 pages. 1986.

Vol. 1228: Multigrid Methods II. Proceedings, 1985. Edited by W. Hackbusch and U. Trottenberg. VI, 336 pages. 1986.

Vol. 1229: O. Bratteli, Derivations, Dissipations and Group Actions on C*-algebras. IV, 277 pages. 1986.

Vol. 1230: Numerical Analysis. Proceedings, 1984. Edited by J.-P. Hennart. X, 234 pages. 1986.

Vol. 1231: E.-U. Gekeler, Drinfeld Modular Curves. XIV, 107 pages. 1986.